Bioethics
Perspectives and Dilemmas Before to Beyond Life

Kusum

2024
Regency Publications
A Division of
Astral International Pvt. Ltd.
New Delhi – 110 002

First Published, 1998

Reprinted, 2024

ISBN: 978-81-8603-080-6 (HB)

Published by : **Regency Publications**
A Division of
Astral International Pvt. Ltd.
– ISO 9001:2015 Certified Company –
4736/23, Ansari Road, Darya Ganj
New Delhi-110 002
Ph. 011-43549197, 23278134
E-mail: info@astralint.com
Website: www.astralint.com

CONTENTS

SECTION V: DAMAGES FOR WRONGFUL BIRTHS

SECTION VI: SEXUAL IDENTITY ABERRATIONS/CONFLICTS AND TRANS-SEXUALISM

SECTION VII: DEATH AND DYING

SECTION VIII: TRADING IN ORGANS/CORPSE

APPENDICES

INTRODUCTION

Bioethics, as defined by the International Association of Bioethics is the study of ethical issues arising out of health care and the biological sciences. The world today is confronted with innumerable dilemmas because of the advancements in medical science, procedures and technologies. The social, legal, moral and ethical issues surrounding these developments are many.

The problems are more serious when these procedures and technologies are used not only for facilitating diagnosis, improving treatment and disease prevention but when they become profit making industries for private practitioners and hospitals. New devices and technologies encourage commercialisation to promote some groups which can benefit from them. Also the more advanced the technology, the greater the costs of health care. Another dimension or fallout of these technologies and procedures is their use for solving not only medical conditions but their application for satisfying individual or social or cultural non-medical needs. Use of procedures for assisted reproduction or sex selection are examples of this.

In a country where even the basic health care needs like safe drinking water, sanitary facilities, and immunisation against infectious diseases is still inadequate, where maternal and infant mortality rates are very high and where malnutrition and even starvation exists, the import and application of certain medical procedures and technologies are bound to raise controversies.

This small book is a collection of articles on various issues in bioethics written over a period of two decades. Since these articles were written at different times for different places, there is repetition at some places which could irk the reader. My apologies for that. It is significant to note that many of these articles were written at a time when these were considered utopian and unrealistic, in the Indian context specially. But they were visualistic. We are very much living in a world where organs of live, brain-dead and cadavers are being transplanted into human bodies, where even pig hearts are being transplanted, where life can be mechanically prolonged through machines, where a pregnant woman with severe brain damage can be kept alive to bring the pregnancy to term[1], where the infertile couples can have children asexually through assisted reproductive technologies, where abnormalities can be detected, and at times, treated pre-natally, where surgical male to female and vice versa carving has become a reality, and much more. A time may not be too far away, perhaps, when synthetic sperms may be available or when males may carry and deliver babies. Though one feels proud about the laudable achievements of medical science — the beneficiaries will be only a small group of priveleged people — the politically resourceful and the rich. For the common man, the five star and the extraordinary treatment procedures are no less than aspiring for a moon.

Some of the broad themes which form the subject matter of the articles included in the book are abortion, sex selection, assisted reproduction, foetal rights, suicide, aberrations and conflicts in sexual identity and allied issues.

Until a few years back it was unimaginable that a child could be procreated, conceived, gestated or delivered in any manner or procedure other than natural heterosexual union and natural child-birth. However, now with newly acquired knowledge, procedures and technologies, wombs and gametes will be purchasable commodities and children artificially produced. There has been a lot of development and advancement in the area all over the world and India has not remained impervious to the new medical procedures and technologies of assisted reproduction — despite economic constraints and the socio-cultural notions which inhibit people to openly discuss any issue related to sexuality.

Several infertility centres have been set up in different

parts of the country where couples are helped to have a baby through various assisted procedures. The most commonly known procedure is artificial insemination (AI). There are many sperm banks which provide sperms for insemination. The procedure recently got significant boost in Madras[2] with the Madras Andrology Research Centre opting to use cryo-preservation techniques for the purpose of freezing and storing donor sperms and using them at a much later date. Infact the popularity of this procedure coupled with good business for professionals, has led to the mushrooming of centres where AI is available. However, as revealed by Dr. R.A. Pai of the Cryogenic Pvt Ltd. (a semen bank) there is a shortage of donors. "We get only about hundred samples every month though the demand is distinctly higher," he says. It is believed that more than a hundred artificial inseminations take place in Delhi every day. In many cases couples seek insemination by relatives or friends, though doctors advise against it. In some cases, the insemination is sought from a husband's relative because of fear of property being inherited by a child in whose veins the family blood does not flow. In one case[3] a gynaecologist in Delhi was persuaded to impregnate a woman with her father-in-law's semen.

In vitro fertilisation procedures are well known at several fertility centres all over the country. According to a recent newspaper report,[4] a test tube baby festival was held at Jaipur where over a hundred children born of embryos developed with IVF techniques assembled. The festival was organised to celebrate the eighth anniversary of the Jaipur Fertility and Micro Surgery Research Centre.

Ovum donations are also known which is evident from the following ad in a leading women's magazine.[5]

> "A young couple, looking for a "female ovum donor". Applicant should be of good complexion and below 30 years of age. Strict confidentiality will be maintained for the applicant. All expenses will be borne by the advertiser and an attractive award for the donor. Please apply with complete particulars at the first instance."
>
> The Times of India Group
> Dr. D.N. Road
> Bombay

According to Dr. Indira Hinduja of the Hinduja Hospital (Mumbai) egg donors ask anywhere between Rs. 10,000 to Rs. 30,000 and doctors are facing a dilemma whether or not to encourage paid egg donors. According to her, two of the human embryos she recently created in the test-tube have been inserted in "surrogates". She did not say whether they were paid surrogates or hiring their space free. (As reported in the *Hindustan Times*, October 29, 1995).

Thus, hiring of wombs is also no longer a strange practice as an assisted reproductive procedure in our country. This is further amply borne by Nirmala Rani's case (Chandigarh) which has accidentally become news.[6]

Apart from these, some more advanced technologies of assisted reproduction are also coming up. The ICSI (Intra Cytoplasmic Sperm Injection) involves the injecting of sperm into egg cell outside woman's body using a specially made pipette. The egg cell is cleaned of other cells around it. The pipette sucks in the head of the sperm after cutting off its tail. The pipette is then pierced into the egg and sperm is released gently. Once this is done, within 18 hours the fertilised egg cell starts multiplying and it is inserted into the uterus of the woman and a normal pregnancy ensues. Such pregnancies have been reported.[7]

TSEICSI (Testicular Sperm Extraction with Intracytoplasmic Sperm Injection) is another reported technique for artificial conception which can help even sterile men to father children. In this procedure, even a sperm may not be required, but just cells from pre-sperm cells. According to Dr. Aniruddha Malpani of the well known Malpani Infertility Clinic at Mumbai, who performed the procedure, in case of azoospermia (absence of spermatozoa) donor insemination was the only treatment which meant that the child born would not be the man's genetically. However, with TSEICSI, sperm was extracted by a testis biopsy and spermatids recovered from the testis were injected into the wife's eggs.[8] Thus we see that many new procedures, medical and non-medical, have come up to assist childless couples to have children. Inspite of the high costs, uncertain success rate and other risks involved in the procedures, desperate couples will continue seeking their help. The Artificial Insemination (Human) Act, 1995, enacted by the Delhi government seeks to regulate certain procedures of assisted reproduction but the risk of clandestine practice and commercialisation ignoring norms of ethics and safety is great. As observed by one writer,[9] "donor insemination has been made into a backyard industry by many malpractising doctors." At one seminar on medical ethics organised by Max Mueller Bhavan, New Delhi and the All India Institute of Medical Sciences in October 1995, Dr. Lalita Badhwar, a Delhi based gynaecologist, pointed out that most sperm banks did not test for HIV. Since semen is one of the most potent means of transmitting the virus, artificial insemination is dangerous. Dr. Indira Hinduja of Mumbai pointed out that her hospital had, in fact, totally stopped artificial insemination because it was morally wrong to use untested donor sperm[9a].

The Indian Council of Medical Research is however working on guidelines to regulate the various procedures and we may soon have a central law to regulate assisted conception and reproduction applicable to the whole country.

Abortion has been legalised by the Medical Termination of Pregnancy Act, 1971 (MTPA) under certain circumstances and conditions. Prior to the Act, the Penal Code permitted abortion only when it was necessitated to save the life of the pregnant woman. Termination of a pregnancy under any other circumstances was an offence. Despite this, however, women used to go in for abortions but it being an offence, the operation used to be performed surreptitiously and under unhygienic conditions and very often by quacks and midwives. This posed a threat to the life and health of the woman. Besides, it was felt that there could be several other situations in which continuation of the pregnancy may not be desirable and so termination should be legally permitted in the interest of the mother, the child and the society. It cannot, however be denied that the demographic needs of the country too must have weighed heavily in the minds of the legislators. Though the right to have an abortion is not absolute, the exceptions are wide and vague enough to permit the same almost on demand. This, coupled with new procedures which enable sex detections, has led to the practice of female foeticide. The idea behind the pre-natal diagnostic procedures was to detect genetic and other abnormalities at a foetal stage but since these tests can also detect the sex of the foetus, these are more often [mis]used only for this purpose. In view of our socio-economic and cultural conditions where sons are preferred to daughters, the 'boy-girl' tests are commonly known. If the test shows a female foetus, the same is often aborted. Only in exceptional cases would one find the use of pre-natal testing procedures for detecting any abnormality in the foetus. When the practice assumed alarming proportions, the government enacted the Pre-natal Diagnostic Techniques (Regulation and Prevention of Misuse) Act, 1994.[10] The Act, however, has not, nor could it be expected to put an end to the practice. Abortion being legal, and the monitoring of the reasons for abortion impossible, couples wanting to get rid of a female foetus can easily get the pregnancy terminated.

An issue which recently came up for debate was the hysterectomy of the mentally handicapped. The controversy was triggered following the incident of mass hysterectomy of mentally retarded women in a state-government-run institution for the mentally deficient girls. Compulsory operations of these girls was justified by some as being appropriate both from human as well as from practical point of view, while others viewed it as barbaric and a clear denial of rights of these girls. According to Dr. S.D. Sharma,[11] Director of the Institute of Human Behaviour and Allied Sciences of Shahdara, Delhi, the test of any civilised society lies in the way it looks after those who cannot look after themselves. "If the argument of maintaining their hygiene is stretched further, then we will say that a person who cannot earn should be hanged and those who are crippled should be killed," he remarked.

Incompetence per se should not deprive any one of advantages that he/she would otherwise have access to. Also, last resorts should not become first options and the sterilisation of the mentally handicapped cannot be ethically justified as a matter of general policy.

Research in foetal tissues is another development in medical technology which poses several legal and ethical dilemmas. Illegal abortions might be encouraged and women hired by pharmaceutical companies to conceive and abort and provide the abortus to them for experimentation and use in the preparation of drugs and cosmetics. According to a recent report,[12] an Eye Institute at Hyderabad was experimenting on eight blind boys to find out if injecting brain cells of aborted foetuses into the retina restored their sight which was lost due to a progressive disease called retinis pigmentosa. The Institute carried out its project in collaboration with some institutions in the United States where such experiments involving foetal cells are not permitted. The Indian Council of Medical Research however, refused to give clearance to this project when it came before it for financial support.

Replacement of diseased and defective organs by healthy ones has been made possible by medical science today. These organs can be obtained from live donors, cadavers or the brain dead. While some organs can be donated with little or no risk to the live donor, there are organs which can be removed from cadavers and transplanted within twenty four hours of the heart ceasing to beat. There is still another kind of transplant where organs have to be removed from heart-beating donors, i.e. donors who are brain-dead but whose heart and lungs are functioning either with the help of mechanical devices or otherwise. A heart beating donor is necessary for organs like the liver, heart, pancreas and lungs as they are very susceptible to lack of blood and their cells die fast. Since the success rate of these operations is going up, the demand for organs is increasing too. In view of the imbalance in the demand-supply ratio, there is always a risk of the poor being hired and exploited. Rich people from other countries come to India for kidney transplants. Rackets in organ trading, 'organ bazars' and 'kidney tours' are being reported from time to time. According to a recent UNI report, the Customs Officers uncovered a 'kidney tour' racket by conducting body searches on people 'suspected of smuggling contraband in pouches sewn into their abdomen'. A group of travellers had suspiciously identical stitches on their bellies but the X-ray and other methods failed to reveal anything incriminatory. The assistance of a doctor as an expert was therefore sought and it was discovered that these people were taken to Latin America for removal of their kidneys. Such 'kidney tours' had been made by hundreds of others before them, the donors/travellers told the doctor. A poor man is prepared to sell his kidney or eye and the middle man arranges everything and profits the most. There are regular reports of the poor, illiterate and gullible being kidnapped, cheated, lured or forced to part with their organs for a nominal price. The practice has gained a lot of notoriety where even medical professionals are alleged to be involved. In order to put a check on exploitation and commercialisation of organs and also to regulate the same to ensure the safety of the donor and donee, the Transplant of Human Organs Act, 1994 has been passed. It makes provisions for live and cadaver donations under specified conditions. In case of live donors, it provides that the donor and recipient should be near relatives. Exceptions however have been provided, where after the approval and permission of the Authorisation Committee set up under the Act, a non related person may donate his/her organ.

In this context there is a recent case from Karnataka[13] where the court allowed transplantation from a non-related donor. The donor was the sister-in-law of the recipient who wished to donate her kidney out of "affection and attachment". The Authorisation Committee had refused permission on the ground, that the donor was not genetically related to the recipient and that the other near relatives were not being considered as donors by the family. On petition to the High Court, permission was granted. The court held that there is no provision in the Act which prohibits a person who is not a "near relative" by definition, from donating his kidney merely because the "near relatives" have not been considered. The court further directed that the Authorisation Committee ascertains from the donor whether she would be donating the kidney "out of affection and attachment."

Considerations like the donor's relationship with the recipient, period of acquaintance and the degree of association, reciprocity of feelings, gratitude and other human bonds were held to be some guidelines to sustain "affection and attachment". Further, a responsibility was cast on the Committee "to ensure that the human organ does not become an article of commerce. The main thrust of the Act is against commercial dealings in human organs."[14]

The working of the Act over a period of few years has revealed its lack of efficacy and other practical problems. First and foremost is the fact that the Act is not applicable throughout the country. The recent incidents of kidney rackets in Noida (U.P.) and the alleged involvement of some Nursing Homes there has created a lot of controversy.[15] The doctors feel that there are several discrepancies in the Act. "The Act is not clear on the issue of unrelated donors. The high powered and notified authorisation committees of the Central Government approves such donors and treating team of doctors are not involved in the process. Still the doctors are liable to face serious problems due to fabricated and false complaints."[16]

It is significant to note that the Act has introduced brain-stem-death as the definition of death. This would spare the relatives a long and agonising bed-side vigil and unnecessary financial burden and also make organ transplantation possible. Organs like heart, liver, pancreas and lung can be successfully transplanted only from a brain-dead person. It is however, important to reflect on the expediency of introducing brain-stem-death as the definition of death in India. It is likely to give rise to strong public apprehension of "premature grave robbery". The illiterate and poor might be exploited and coerced and their seriously ill relatives declared brain dead so as to facilitate organ retrieval. There could also be an error of judgement. In our context where the level of literacy, vigilance and awareness is low, these fears are more real.

The sustenance of the brain dead and terminally ill patients pose difficult situations and dilemmas in the face of advanced medical procedures and technologies. Mechanical prolongation of heart beating is now possible by the use of respirators, ventilators and other gadgets. But, "What he is living with no longer was bestowed by nature or God, but was a special existence created artificially by high technology."[17] This gives rise to a series of medical, and ethical issues. Who is to decide whether to continue or withhold medical support to such patients? How far should one go in prolonging the life of a terminally ill patient? Is it worthwhile to spend huge amounts of money only to prolong the dying process without reversing the fatal conditions of the patient? In trying to prolong such life by a few days or few weeks or few months, one may use up vital resources which otherwise could be used to treat those who have better chances of recovery. Moreover, prolongation of life without regard to the quality of life may not infact be in the best interests of the patient. Infact in many cases it may not be compatible with respect for dignity of human life. Whatever it is, the situations pose difficult choices for the treating physicians, and conflict and dilemmas for the relatives.

Closely connected with this is the issue of euthanasia,[18] or mercy killing as it is commonly known. It is a deliberate bringing about a gentle and easy death making the last days of the patient as comfortable as possible. Euthanasia could be voluntary (when requested by the patient) or involuntary or compulsory (when it is without consent) also, it could be passive, i.e. by disconnecting life support systems, or active, i.e., when something is administered to the patient to hasten the dying process. There is a lot of debate on this subject with strong pleas to legalise it on compassionate grounds. In a majority of countries euthanasia is considered to be a crime but, some countries like Holland, have legalised it. The judgement of the Supreme Court in *Gian Kaur* v. *State of Punjab*,[19] is significant in this context. Apart from over ruling *P. Rathinam* v. *Union of India*,[20] which decriminalised attempt to suicide, the court also discussed the issue of euthanasia. It was argued by the supporters of euthanasia that the "right to die" with dignity should be read into "right to live with dignity" when death due to termination of natural life is certain and imminent and the process of natural death has commenced. In such cases, it was argued, it was not extinguishing life but only accelerating conclusion of the process of natural death. The court, however, declined to accept this argument and remarked:[21]

> It is sufficient to reiterate that the argument to support the view of permitting termination of life in such cases to reduce the period of suffering during the process of certain natural death is not available to interpret Article 21 to include therein the right to curtail the natural span of life.

However, though active euthanasia could constitute the crime of homicide, passive euthanasia is well known as good medical practice being in the best interests of the patient.

Decriminalisation[22] of attempt to commit suicide has since long been an issue of academic and judicial debate. The judgement of the Supreme Court in *P. Rathinam*[23] declaring section 309 as ultra vires the Constitution was hailed as a landmark judgement. The verdict, however, was short-lived and was set aside in *Gian Kaur.*[24] The court in this case, held that the "right to life" under article 21 of the Constitution does not include the "right to die". Thus attempt to commit suicide continues to be an offence. It would indeed have been a dangerous proposition to provide a blanket protective shield to all cases of attempted suicide. What is significant infact is humanising of the judicial approach rather than humanising a law by deleting altogether a provision without regard to subjectivity.[25]

This is the journey of the book, from before to beyond life, full of excitements and dilemmas created and compounded by the marvels of science which at times seek to play God and at times end up playing havoc. Balancing the import and application of these technologies and procedures with wisdom, keeping in mind the socio-cultural and economic conditions of a society is the clarion call.

REFERENCES

1. *R.V. Potter* was a case of "reanimation donation" from the United Kingdom where a victim of assault who died of head injuries was put on ventilation *for the specific purpose of removal of kidneys for transplantation.* The accused however, got the advantage of the defence of *novus actus interveniens* as he ascribed death not to his attack but to the operation. [see Mason and McCall Smith, *Law and Medical Ethics* (1983) at 187]
2. See Rasheed Bhagat, "Misconceptions of Conception", *Indian Express*, September 17, 1994.
3. Reported in Durba Banerjee, "The Baby Factories," *Statesman*, June 2, 1996.
4. "Test Tube Baby Festival held at Jaipur". *Observer*, May 1997.
5. See Ziya-us-Salam, "What Price for an Ovum", *Statesman*, June 2, 1996.
6. Nilanjan Mukhopadhayay, "Nirmala's Katha", *The Hindustan Times*, (Sunday) June 29, 1997
7. See, "Novel fertilisation process raises new hope", *Hindustan Times*, August 21, 1997; See also, "Baby concieved by new technique", *Indian Express*, May 15, 1997.
8. See, Shameen Akhtar, "Goodbye to Sperm Banks", *Telegraph* May 20, 1996; See also "No sperm? A spermatid will do", *The Independent* (Bombay) April 13, 1996.

9. See, Durba Banerjee, *Supra* note 3.
9a. See, *Issues in Medical Ethics,* vol. 5, No. 2, April–June 1997 at 53–54.
10. The state of Maharashtra passed the Maharashtra Regulation of Pre-natal Diagnostic Techniques Act in 1988. Since it was a state legislation, the practice continued with impunity in other parts of the country. Hence the need for the central law.
11. See "Hysterectomy on Retarded" by Kalpana Jain, *Times of India,* February 22, 1994.
12. PTI Report. "Eye Institute experiments on blind boys with grisly cocktail", New Delhi, December 26, 1996, reported in *Issues in Medical Ethics,* vol. 5, No. 2, April–June 1997 at 64.
13. *B.L. Nangia* v. *Kantha,* AIR 1996 Karn. 82.
14. Id., at 83.
15. See "Nephrologists to boycott dialysis work today", *The Hindustan Times,* June 3, 1998, p. 1.
16. Dr. D.S. Rana, Secretary, Delhi Hephrology Society, as reported in the *Hindustan Times,* id.
17. Prof. Ai, Chinese physician clinical ethicist, Centre for Medical Ethics, Ruhr University Bochun, New Case Studies for Bio-medical diagnosis, pp. 9–10, quoted by Reng-Zong Qui in "Bioethics in Developing Countries", in *Bioethics,* vol. 7, No. 2–3, April 1993 at 112.
18. The word is derived from the Greek words "eu"= good and "Thanatos" = death meaning thereby, good death.
19. 1996 Cri. L.J. 1660 (Supreme Court).
20. 1994(3) SCC, 394.
21. *Supra* note 19 at 1666.
22. Section 309 of the Indian Penal Code, 1860 makes attempt to commit suicide a punishable offence.
23. *Supra* note 20.
24. *Supra* note 19.
25. See, Kusum (ed.) *Suicide: Some Reflections* (1995), Regency Publications, New Delhi.

SECTION I

ASSISTED REPRODUCTION

CHAPTER 1

ARTIFICIAL INSEMINATION AND THE LAW*

Procreation is a natural aspiration of every human being. The advancement in science and technology has not only enabled mankind to restrict and control birth rate by the use of contraceptives but has also enabled childless couples to become parents through the device of artificial insemination, i.e., transmission of male seeds into the female body artificially through a syringe.

The history of artificial insemination can be traced back to 1322 when mares were successfully breeded by the Arabs. The superior mares of the enemies were artificially impregnated by the sperms of inferior horses to weaken them. From this time onward, until the middle of the sixteenth century, successful experiments were made on fish and other animals. The first successful insemination of a human being is credited to Dr. John Hunter, an English physician, who used this device upon a married woman using her husband's semen in the late eighteenth century. In 1890, Dr. Robert L. Dickinson established the practice of artificial insemination using a donor's semen. However, it was not until the twentieth century that recourse was had to artificial insemination for the purpose of procreation.[1]

---

* 19 Journal of Indian Law Institute, 283 95 (1977)

Artificial insemination can be practised in three ways. First is the artificial insemination homologous or husband (AIH) where the semen injected into the female body is that of her husband. This is resorted to when some weakness due to physical or psychological reasons prevents copulation or when the husband or wife is impotent (not sterile). In the second type, semen is that of a third party donor and is known as artificial insemination donor (AID). This is resorted to, normally, when the husband is sterile or suffering from some hereditary disease, etc. The third kind of artificial insemination, which is not very popular, is one in which the seed of the husband and that of a third party is co-mingled. This is known as confused or combination artificial insemination (CAI).

Though not popular as yet in India, the practice of artificial insemination is bound to attract the attention of medical practitioners and childless couples, and give rise to legal problems. Its use is on the increase in many western countries, where it has given rise to several legal questions. Does artificial insemination amount to adultery on the part of the wife? Does it amount to consummation of the marriage? Is a child born by this device a legitimate child? What are the custodial and visitation rights of the father? Is the father bound to maintain such a child? And so on. It may be pointed out here that in the case of AIH no question of wife's adultery or child's illegitimacy arises since the seed that gives birth to a child is that of the woman's legally wedded husband. Problems arise when the donor is a third party.

Adultery

The word adultery has been derived from the Latin term *adulterare* which according to Harper's Latin dictionary means, 'to defile' 'to counterfeit', 'to adulterate'. Introduction of something spurious is adulteration. Since AID has the potential of introducing into the husband's family a 'spurious' child, it has been argued by some that it (AID) amounts to adultery. The question that arises is, what constitutes adultery? Is sexual intercourse a necessary ingredient of adultery? What constitutes sexual intercourse? Is penetration a necessary ingredient? An so on. The ancient Indian law-givers recognised three degrees of adultery. A man meeting a woman in a lonely place at an improper time talking to her, casting amorous glances, smiling at her, are examples of adultery of the first degree. Sending messages to a woman, touching her ornament or clothes also constitute adultery of the first degree according to Vyasa. Sending perfumes, garlands, fruits, wines, food or clothes and conversing with her in secrecy is adultery of the second degree.

Sitting on the same bed, dallying, kissing and embracing, this is adultery of the highest degree.[2] These, however, are outmoded concepts and modern definitions require sexual intercourse between two persons of the opposite sex, at least one of whom is married, as a necessary ingredient of adultery.

The Indian Penal Code defines adultery as:

> Whoever has sexual intercourse with a person who is and whom he knows or has reason to believe to be the wife of another man, without the consent or connivance of that man, such sexual intercourse not amounting to the offence of rape, is guilty of the offence of adultery . . . .[3]

Thus, the essential ingredients of adultery are:

(i) The accused must have sexual intercourse with a woman.
(ii) That woman must be the wife of another person.
(iii) The accused must have known or should have reason to believe that the other party was the wife of another person.
(iv) The sexual intercourse should not amount to rape.
(v) The act must have been performed without the consent or connivance of the husband.

Sexual intercourse thus is a necessary ingredient of the offence of adultery. Even under the Hindu Marriage Act, 1955, in the absence of any special definition, "adultery means consensual sexual intercourse between married person and another person of the opposite sex during the subsistence of the marriage."[4]

What constitutes sexual intercourse is another question. Under the explanation to section 375 of the Indian Penal Code, which defines rape, mere penetration is sufficient to constitute the sexual intercourse necessary to the offence of rape. The same test may be applied in the case of adultery as well. There have been, however, cases under the English law where, even in the absence of an act of sexual intercourse, a verdict of adultery was given. In *Rutherford* v. *Richardson*[5] and *Russell* v. *Russell*[6] adultery was said to have been committed even though there was no penetration but "some lesser act of sexual gratification." In *Sapsford* v. *Sapsford and Furtado*[7] the court recognised that voluntary submission to or participation in intimate physical contacts may be inconsistent with the duties of a wife towards her husband. Karminiski, J.,[8] stated: "I have no doubt that, if whole penetration was not achieved, some lesser act of sexual intercourse was performed", and granted a divorce on proof of such act. Though these cases had nothing to do with artificial insemination, yet the effect is the demolition of the concept of sexual intercourse with penetration as an ingredient of adultery.

The Indian position, however, continues unchanged and penetration whatever the extent, irrespective of ejaculation is a necessary ingredient. In *W.J. Phillips* v. *Emperor*[9] the court held that "sexual intercourse is a necessary ingredient of the offence under S. 497, I.P.C. Nothing short of it would justify a conviction under that section."[10] Similarly, In *re Anthony*[11] where a boy of fifteen years was accused of committing rape on a girl of seven and half years, the court held that "while there must be penetration in the technical sense, the slightest penetration would be sufficient, and a completed act of sexual intercourse is not at all necessary."[12]

The issue whether artificial insemination amounts to adultery was raised for the first time in a Canadian court in *Orford* v. *Orford.*[13] The marriage was not consummated because of some physical affliction of the wife. The wife, however, gave birth to a child. The husband accused the wife of adultery, while he stayed in Canada and she was in England. According to the wife, she had resorted to artificial insemination as the doctors had advised her that the only way she could be cured of the affliction was by bearing a child and that it might be done artificially. The court disbelieved the story and came to the finding that the wife had sexual intercourse in the ordinary way. It, however, dealt with the aspect of artificial insemination also and observed:

> That no authority can be found declaring, directly or indirectly, that "artificial insemination" would constitute adultery is not to be wondered at. This is probably the first time in history that such a suggestion has been put forward in a Court of justice. But can any one read the Mosaic law against those sins, which, whether of adultery or otherwise, in any way affect the sanctity of the reproductive functions of the people of Israel, without being convinced that, had such a thing as "artificial insemination" entered the mind of the lawgiver, it would have been regarded with the utmost horror and detestation as an invasion of the most sacred of the marital rights of husband and wife, and have been the subject of severest penalties?[14]

The court continued further:

> In my judgement, the essence of the offence of adultery consists, not in the moral turpitude of the act of sexual intercourse, but in the voluntary surrender to another person of the reproductive powers or faculties of the guilty person; and any submission of those powers to the service or enjoyment of any person other than the husband or the wife comes within the definition of "adultery."
>
> ... So long as nothing takes place which can by any possibility effect that function, there can be no adultery; so that, unless and until there is actual sexual intercourse, there can be no adultery. But to argue, from that, that adultery necessarily begins and ends there, is utterly fallacious. Sexual intercourse is adulterous because in the case of the woman it involves the possibility of introducing into the family of the husband a false strain of blood. Any act on the part of the wife which does that would, therefore, be adulterous. That such a thing could be accomplished in any other than the natural manner probably never entered the heads of those who considered the question before.[15]

The question whether AID amounted to adultery or not was discussed at length in the House of Lords during its debates in March 1949.[16] "Adultery", according to the Lord Archbishop of Canterbury,

> is the surrender outside the bonds of wedlock, and in violation of it, either of the sexual organs alone . . . or of the reproductive organs alone by AID or, of course, of both, as in normal intercourse. If that be so, AID is adultery.[17]

Lord Merriman,[18] however, was of the view that penetration of the female sexual organ was an essential element to constitute adultery. According to him Lord Dunedin in his *dictum* in *Russell* v. *Russell*[19] used the phrase fecundation *ab extra* which had nothing to do with artificial insemination. Besides, according to Lord Merriman, AI does not involve the act of sexual intercourse at all and hence no adultery. A penetrating study was made by Tallin who, in an article,[20] argued that AID is adultery. According to him:

> Sexual intercourse without risk of pregnancy, and risk of pregnancy without sexual intercourse, may be equally destructive of a happy marriage relationship, and it may therefore be that either would constitute adultery . . . .[21]

He continues that:

> "If there must be one test and one test only for adultery, it would seem that it should be whether the conduct in question involves risk of pregnancy."[22]

Refuting Tallin's contention that AI constitutes adultery, Hubbard, in his article,[23] points out that:

> The act of sexual intercourse requires the mutual co-operation of a man and a woman acting together at the same time. Their mutual activity constitutes a single act of intercourse. An act must be localised in time. . . . When can the act of "sexual intercourse" be said to have occurred in artificial insemination?. . . . The act . . . cannot rationally be said to have commenced with the ejaculation of the semen and subsisted *qua act* until the insemination took place. This would not be one act but two separate acts.

> The donor and the ultimate recipient of his seed cannot be said to have acted together at the same time in their act. . . . If the activity is discontinued the act ceases. In artificial insemination the act of obtaining the seed is quite distinct from the act of artificial insemination. If a man manufactures a bullet knowing that it will be fired into something, he cannot be said to have pulled the trigger of the weapon and fired the bullet home simply because but for him there would be no bullet.[24]

Tallin, in reply to the argument[25] agreed that an act must take place in time "but it does not follow that it must be instantaneous." According to him:

> In the practical affairs of life with which courts have to deal it may only be necessary to say that an act began not later than a certain moment and ended not later than another certain moment. An act of murder may be held to have begun not later than the purchase by the murderer of the poison and ended not later than the moment the victim was pronounced dead by the doctor. It is not necessary, in order to find that murder or an act of murder has been committed, to say that it took place when the poison was put in the victim's coffee or when he drank it, or when he died. . . . Each had a beginning, duration and ending which might defy definition.[26]

According to Tallin, it would be equally impossible to determine the time of sexual intercourse because of the various processes involved in it.

The question came up for consideration comparatively recently in the Scottish Court of Session at Edinburgh in *Maclennan* v. *Maclennan*.[27] The husband filed a petition for divorce on the ground of his wife's adultery. The wife alleged that the child she gave birth to, was conceived by artificial insemination. The husband contended that AID was adultery in the eye of law and also that he did not consent to his wife's impregnation. The court, however, conceded that a married woman committed a grave and heinous offence against the marriage contract by submitting to artificial insemination without her husband's consent, nevertheless, this, according to the court, was a matter in respect of which the legislature should determine an appropriate remedy. It was considered to be quite irrelevant to the issue before the court. In a preliminary judgement, the court held that AID would not constitute adultery in law but adjourned the case for verification of AID by the wife. The wife declined to provide the necessary information and the court held that she had committed adultery. After considering a few English cases,[28] the court derived the following propositions, *viz.*:

(1) For adultery to be committed there must be two parties physically present and engaging in the sexual act at the same time.
(2) To constitute the sexual act there must be an act of union involving some degree of penetration of the female organ by the male organ.
(3) It is not a necessary concomitant of adultery that male seed should be deposited in the female's ovum.
(4) The placing of the male seed in the female ovum need not necessarily result from the sexual act, and if it does not, but is placed there by another means, there is no sexual intercourse.

The court also considered that if AID were deemed to be adultery, the question would arise whether the donor whose seed has been used is also guilty of adultery. If that be so, at what point of time is adultery committed? If adultery is deemed to take place at the time of the parting of the seed then, suppose the seed is never used? If the crucial time is the time when the seed is injected into the woman's body, then suppose, this is done after the donor's death? (Since the seed can be preserved for some time). In that case the woman's conduct would constitute not only adultery but necrophilism. Such a proposition according to the court would be absurd. In view of all this, it found that artificial insemination by donor does not constitute adultery.

The decision of Lord Wheatly in the above case was the subject of debate in the House of Lords in February 1958[29] as a result of which a departmental committee was appointed in September to enquire whether, and if so, what changes in the law were necessary, in view of this practice. The committee expressed the view that a clear distinction be drawn between artificial insemination and adultery. It endorsed the view of the Royal Commission on Marriage and Divorce 1951–1955 that artificial insemination of the wife without the consent of her husband be made a new and separate ground of divorce or judicial separation. It may be pointed out here that from a physician's point of view also the two, *viz.*, adultery and artificial insemination are different. The object of an adulterous act is carnal pleasure and emotional and physical satisfaction. These are missing in artificial insemination which is resorted to for the purpose of begetting offspring. In fact, they are antithesis of each other. Whereas, in artificial insemination the only purpose is procreation, *i.e.*, it is not a sexual but a professional and therapeutic act, in adultery it is sensual pleasure alone and if pregnancy results, it is normally unsolicited. The reason why adultery has been made an offence or a ground for divorce appears to be the risk of pregnancy involved in the act. Besides, it disturbs marital peace and is an invasion of the right of a spouse to have exclusive sexual pleasure from the other. The Indian law, as stated above, when it talks of merely penetration (without ejaculation) does not confine adultery to a situation which might lead to conception. However, in the present context, suffice it is to note that carnal knowledge with the risk of pregnancy will certainly constitute adultery, though risk of pregnancy alone cannot be the basis of adultery. In the present day scientific development when resort to contraceptive devices and sterilisations are increasingly common and popular, the risk of pregnancy is greatly diminished or eliminated even in a complete normal sexual act. In artificial insemination the basic question is whether lack of carnal knowledge in the actuality of conception or introduction of a foreign element with a view to conception would amount to adultery or not. It seems that in the absence of this element, the answer should be in the negative.

One argument that is often given in support of AID is that by the use of this method, the marriage bond is violated voluntarily, for a good reason, by the consent and wish of the husband and wife concerned. A childless marriage often causes severe pyschological disturbances bringing in its wake, strain, stress and rupture. Artificial insemination if successful, comes to the rescue of the couple. Its opponents, however, argue that instead of solving the difficulties, AID might create more problems. According to Lord Archbishop of Canterbury:

> [A] husband often has a great longing to have children, and sometimes his longing is frustrated by the sterility of his wife. No one has yet suggested that he should relieve his frustrations by introducing into his family a child whom he has fathered by AID or otherwise by another woman. He has to bear his frustration, and so can a wife.[30]

He continues:

> If difficulties and jealousies arise in their married life these will surely be complicated and aggravated by the father's knowledge that the child of the marriage is not his at all. It is not to be supposed that every marriage will be delivered from all frictions from the fact that AID has been employed and the intruded child may easily become a source of recrimination.
>
> . . . In the interests of society, of the family, and of the child . . . I think it should be neither encouraged nor permitted.[31]

The practice has been denounced on spiritual grounds by Lord Archbishop of Canterbury who, quoting the words from an article appearing in the February issue of the *Fortnightly*, says:

> It is the fundamental principle of Christian belief that God is Love and our common Father, from which belief springs the dignity of human personality. This requires its corelative principle that children should be born of love of their parents. In AID the purely material and mechanical element in procreation is separated, and not temporarily but absolutely separated, from the organic and personal lives of the two persons involved. . . . In AID human life is initiated as the result of a momentary mechanical process entirely divorced from the spiritual, mental, emotional and physical lives of its parents. If extensively practised it would inevitably degrade the whole conception of

> personality and intensify that tendency towards the reduction of life to mechanism which is one of the most sinister features of our time.[32]

The foregoing discussions thus reveal that there are both antagonists as well as protagonists of the practice of artificial insemination. In the absence of consensual intercourse and penetration it should not be deemed to constitute adultery for purposes of criminal law. Nevertheless, it cannot be denied that it is bound to have far-reaching effects on family life. A child so conceived, especially without the husband's consent, will in all probability disrupt marital peace and harmony. On the other hand, the practice is likely to be preferable than adoption, since an adopted child is a hundred per cent stranger whereas a child artificially conceived by the wife would be at least fifty per cent the couple's own child as the mother would be the husband's own wife. In view of this, there is a definite need for legislative provisions to regulate and control the practice. While it may not constitute adultery, it may well be made a ground for matrimonial relief for the husband in case the wife resorts to insemination without the husband's consent.

Legitimacy and Custody of the Child

Another controversial issue that an AID child poses is whether such a child is the legitimate child of the father and as to the father's right and duties *vis a vis* the child. In *Strnad* v. *Strnad*[33] the question of custody of AID child was considered by the New York Supreme Court. It was held that the husband had a right of visitation although the child was not his offspring biologically. The court observed:

> The child has been potentially adopted or semi-adopted by the defendant. In any event in so far as this defendant is concerned, and with particular reference to visitation, he is entitled to the same rights as those acquired by a foster parent who has formally adopted a child, if not the same rights as those to which a natural parent under the circumstances would be entitled. . . . In the opinion of this court. . . . this child is not an illegitimate child.[34]

In *Doornbas* v. *Doornbas*,[35] however, the court took a different view. It held that AID was adultery and contrary to public policy and that the child born through this process was illegitimate. It further ruled that although the husband had consented to artificial donor insemination of his wife, yet he had no visitation rights to the child. The court said:

> Heterologous artificial insemination (when the specimen of semen used is obtained from a third party or donor), with or without the consent of the husband, is contrary to public policy and good morals, and constitutes adultery on the part of the mother. A child so conceived is not a child born in wedlock and therefore illegitimate. And as such it is the child of the mother and the father has no right or interest in said child.[36]

In a later case, *Gursky* v *Gursky*[37] again, the New York Supreme Court found that a child born through AID is not a legitimate child of the husband. In this case, the wife, defendant in her husband's unsuccessful action for annulment, counterclaimed for an annulment of the marriage for impotency of her husband and for custody of and support for a child born through artificial insemination with semen from a third party donor. The husband had given his written consent for the insemination, nevertheless, the court granting annulment refused to recognise the child as husband's legitimate child. The effect of the written agreement by the husband, however, was that he was bound to support the child. Similarly in *Anonymous* v. *Anonymous*[38] a husband had consented to his wife's therapeutic impregnation. In a claim by the wife for alimony the husband asserted that such a child was illegitimate. The court refused to accept this argument and awarding alimony held "that the written consent carried an *implied promise* to furnish support for the resulting progeny."[39]

In *People* v. *Sorensen*[40] a divorced woman sued her former husband for non-support of their AID son. In this case, the defendant had consented, after fifteen years of marriage and a medical determination of his sterility, to allow his wife to be artificially inseminated. AID was administered and a child was born. For about four years prior to their separation the defendant represented that he was the child's father. The

California Supreme Court held that the defendant was the lawful father of the child born to his former wife, that the child was conceived by artificial insemination to which the defendant had consented, and that his conduct carried with it an obligation to support.

Thus, we find that there is no unanimity on the point and there are conflicting and contradictory decisions. The author, however, feels that innocent children should not be made to suffer the stigma of illegitimacy. A father who consents to the wife being so impregnated or holds such a child as his own should be treated as the real father of the child and the *Sorensen* decision lends support to this point of view. In case of insemination without consent, the father should not be burdened with the liability to maintain the child. It is necessary that formal written consent should be made obligatory in order to avoid a controversial situation.

Consummation

Another legal issue that stems from the practice of AID is whether AID is tantamount to consummation of marriage, in a case when a wife resorts to AID (whether it is successful or not is a separate issue) and later seeks to avoid the marriage on the ground of non-consummation or impotency of the husband. According to Lord Archbishop of Canterbury, "in marriage, mutual surrender either of the sexual organs without procreation, or of the reproductive organs by means of A.I.H. is to be regarded, in my view, as consummation."[41]

In *R.E.L.* v. *E.L.*[42] a wife whose husband was unable to consummate the marriage because of psychological infirmity was artificially inseminated with her husband's seed. This process spread over a period of twelve months. A few weeks after an insemination, the wife, unaware that she was pregnant, left the husband. The child was subsequently born. The wife filed a suit for nullity of the marriage. The court held that the conduct of the wife in allowing herself to be artificially inseminated with her husband's seed and the conception of the child did not necessarily amount to an approbation of the marriage; nor was the court prevented on any ground of public policy from pronouncing a decree even though the result of that decree would be to bastardize the child. The court observed: "If the child should be made illegitimate it is most regrettable, but the stigma of birth are of less effect that they were, and sons are not judged by the errors of their parents."[43]

The author, however, is unable to reconcile with this point of view. When both the parties had agreed to the insemination and especially when the seed is that of the husband there is no reason why the parties should not have been estopped to plead non-consummation. Besides, it is very unfair that a child who is in every manner legitimate being the biological offspring of both the parties and born within wedlock — should be labelled illegitimate by annulling such marriage on ground of non-consummation.

The issue whether artificial insemination amounted to consummation or not was raised again *Slater* v. *Slater*.[44] In this case AID was resorted to but was unsuccessful and ultimately, a child was adopted. It was held that the attempted insemination could not constitute consummation and that the doctrine of approbation could not be applied to defeat the wife's claim despite the insemination and adoption. However, it all depends on the circumstances of each case and nothing can be said categorically as to whether the act of artificial insemination — whether AID or A.I.H. — would or would not amount to approbation of the marriage. By the very nature of the process of the device, in the absence of organic intercourse, it would be hard to believe that such a marriage is physically consummated, nevertheless, if such a marriage is easily liable to be annulled on the ground of non-consummation, then the very purpose and object of the insemination would be defeated, *viz.*, the purpose of relieving the parties of the frustration of not bearing a child and removing a principal cause of disharmony in the family. In a case where the husband consents and a child is born by this device, it is reasonable and logical to assume that the marriage has been consummated and neither the husband nor the wife should be allowed to plea non-consummation and get out of the wedlock. Difficulty, however, is bound to arise in cases where a wife gets inseminated without the consent of the husband. It appears that in such cases the wife's resort to AID should not deprive the husband of the plea of non-consummation of the marriage.

Conclusion

Thus, we find there is a haze of confusion and uncertainty surrounding the concept of AID. In view of its far-reaching consequences, an immediate legislative action is required. AID should not constitute a criminal act of adultery because of the absence of physical act of union of flesh and flesh. As pointed out by a writer:

> There would be. . . . many problems in the practice of artificial insemination and there would clearly be anomalies arising from the application of the present rules of law until corrective legislation is passed. But let us not complicate needlessly existing concepts in order to find a complete solution to a problem arising from an unnatural practice. To do that would be to revive in the mind of the man in the street his old feeling that "the law's an ass." He can understand that a child born of a wife who has not committed adultery may be illegitimate but not that artificial insemination can be regarded as "fun in bed."[45]

Besides, there is no *mens rea* or guilty intent of the wife resorting to AID. Also, if it were to be treated as adultery, it would give rise to a number of absurdities and complications, *e.g.*, the donor would be an adulteror, the carrier of the semen and the doctor who performs the operation would be participants in the act. Also suppose the seed is used after the donor's death, would the woman be guilty of committing adultery with a dead man? Suppose the doctor performing the operation were a female doctor, what would be the consequence? A very queer situation would arise in a case where the husband himself transmits by natural act a third person's seed to his wife.[46*].

* A new medical technique makes it possible to inject a sterile or sub-sterile husband with a donor's semen so that he transmits it by natural intercourse to his wife.

Nevertheless, it cannot de denied that the practice of AID is very likely to effect family peace and harmony. The presence of another man's child is bound to create emotional conflicts and tensions in the family and, therefore, it would be reasonable, nay necessary, to provide matrimonial relief to the husband if the wife resorts to AID without his consent. Only childless couples where the husband's sterility has been medically established should resort to AID. The written consent of the husband should be taken and if he consents or ratifies the act the child should be deemed to be his legitimate child. Only licenced doctors of high repute should be authorised to perform the operation. The doctor should carefully select the donor and also maintain a strict secrecy. In other words, the needs of a progressive society should be reconciled with the demands of propriety and legal precedent while dealing with such cases.

REFERENCES

1. John W. Hager, Artificial Insemination: Some Practical Considerations for Effective Counseling, 39 *North Carolina Law Review* 217 at 219–20 (1960-61).
2. Ganganath Jha, 1 *Hindu Law and its Sources*, chapter XVIII.
3. S. 497.
4. *Gitabai* v. *Fattoo*, A.I.R. 1966 M.P. 130.
5. [1923] A.C. 1.
6. [1924] A.C. 687.
7. [1954] 2 All E.R. 373.
8. Id. at 375.
9. A.I.R. 1939 Oudh 506.
10. Id. at 509.
11. A.I.R 1960 Mad. 308.
12. Id. at 311, per Anantanarayanan, J.
13. 58 D.L.R. 251 (1921); 49 *Ontario I.R.* 15.
14. Id. at 257–58.
15. Id at 258.
16. *Parliamentary Debates* (*Lords*), Vol. 161: 404 (1948-49).
17. Id. at 404.
18. Id. at 412.
19. *Supra* note 6.
20. G.P.R. Tallin, Artificial Insemination, 34 *The Canadian Bar Review* (1956).
21. Id. at 22.
22. Ibid.

23. H.A. Hubbard, Artificial Insemination: A Reply to Dean Tallin, 34 *The Canadian Bar Review* 425 (1956).
24. Id. at 433.
25. C. Tallin, Correspondence: Artificial Insemination, id. at 628.
26. Id. at 629.
27. (1958) Sess. Cas. 105, cited by E.I. Klayman, Therapeutic Impregnation, 39 *Cincinnati Law Review* 291 at 310 (1970).
28. *Russell* v. *Russell*, *supra* note 6; *Rutherford* v. *Richardson*, *supra* note 5; *Sapsford* v. *Sapsford*, *supra* note 7; *Dennis* v. *Dennis*, (1955) Probate Division 153; *R.E.L.* v. L., (1949) Probate Division 211; *Clarke* v. *Clarke*, (1943) 2 All E.R. 540.
29. 207 *Parliamentary Debates* (*Lords*) 926–1016 (1957–58).
30. *Supra* note 16 at 406–407.
31. Id. at 407.
32. Id. at 408–409.
33. 78 N.Y.S. 2d 390 (1948); the case has been referred to in 21 *Modern Law Review* 236 at 240 (1958).
34. Quoted by G.W. Bartholomew, Legal Implications of Artificial Insemination, 21 *Modern Law Review* 236 at 240 (1958).
35. *Time*, December 27, 1954, p. 41, the case has been reviewed in 41 *A.B.A.J.* 263 (1955); see also 23 *U.S.L.W.* 2308.
36. 41 *A.B.A.J.* 263 (1955).
37. 39 Misc. 2d 1083; 242 N.Y.S. 2d 406 (Sup. Ct. 1963). See also, 39 *Cincinnati Law Review*, *supra* note 27 at 311.
38. 41 Misc. 2d 886; 246 N.Y.S. 2d 835 (Sup. Ct. 1964), referred to in *Cincinnati Law Review*, *supra* note 27 at 316.
39. Ibid.
40. 68 Cal. 2d 280; cited by B.J. Davies, Family Law — Recent cases, 23 *Buffalo Law Review* 548 at 550 (1973-74).
41. *Supra* note 16 at 404.
42. *Supra* note 28.
43. Id. at 219.
44. (1953) Probate Division 235.
45. "*Sed Quaere*", 28 *Australian Law Journal* 612 (1954–55).
46. See also H.R. Hahlo, Some Legal Aspects of Human Artificial Insemination, LXXIV *South African Law Journal* 167 at 173 (1957).

CHAPTER 2

WOMBS TO-LET*

A new company "Surrogate Mothering Ltd." expects to deliver 12 of its products — babies — soon. The announcement came amidst reports that actor Richard Burton had advertised in England for a surrogate on contingency fees. He offered $25,000 for a girl and $50,000 for a boy!

The device of hiring another woman to mother a child has in fact been picking up in recent years, especially in the United States. And one Kentucky attorney has in partnership with a physician formed a firm known as "Surrogate Parenting Associates". There have also been reports of surrogates refusing to part with their babies. In one case a surrogate who was about to deliver a baby to a New York couple changed her mind and decided to keep the baby. In another case, reported from England, a prostitute hired by a childless couple agreed to bear and deliver a child. But when she gave birth to a male child she refused to part with it. The couple went to court but lost.

Though the concept has yet to gain ground in our country it would nonetheless not be premature to reflect on its implications and ramifications. In fact, legal learning should be forearmed with proper guidelines before the issue actually comes up for deliberation so that the courts do not find themselves in a dilemma.

---

* *Hindustan Times*, May 27, 1984

A popular Hindi movie "Doosri Dulhan" has already touched upon this theme and with more and more people coming to know of this device, it is bound to be made use of. Once resorted to, implications and disputes will follow as a necessary corollary. Guidelines for this can be sought from case law under artificial insemination (in other countries) and from the experience of some countries where surrogate mothering is known.

The use of this technique has become an alternative to normal reproduction for many reasons. First and foremost is the inability of the couple to produce a child because the wife may be unable to conceive or carry the fetus to term. There may also be other medical or eugenic reasons like genetic defects, inheritable disease, etc. The use of a surrogate is attractive as a matter of convenience. A wife with a career may not wish to take time off from her pursuits to carry a foetus through. Besides, some women might suffer from anxiety or labor phobia and therefore may not want to carry a child themselves.

Surrogate mothering involves the hiring of another woman who is the surrogate. She is artificially inseminated with the husband's sperm. The process might have to be repeated a couple of times before the surrogate conceives. Upon birth, the child is handed over to the couple. In this case the surrogate is both the genetic (as the ovum is her's) and gestational (as she carry's the child throughout the period of gestation) mother. Where the wife's ovum is fertilised with her husband's semen and then implanted in a surrogate the wife is the genetic mother and the surrogate is the gestational mother.

Until recently a childless couple had to resort to adoption to have a child. A definite advantage of surrogate mothership over adoption is that in this case the child is, at least genetically, the child of the husband and in case of in vitro fertilisation, of both. Also, the availability of children for adoption has become scarce and couples who wish to adopt have to wait a long time. Apart from all this there is always the fear of rejection and emotional conflict should the child come to know that the parents are not his biological parents.

Under the contract made in countries where the practice is known, the couple are the legal parents of the child and the surrogate has to give the custody of the child to them upon birth. The husband being the natural father need not adopt, but the wife has to adopt and the surrogate has to give in adoption.

In addition to reimbursement expenses like diet, medicine, delivery expenses etc., the surrogate is usually paid a fee around $10,000. The contract may moreover require the surrogate not to smoke, drink or take drugs during the period. Abortion may be specified in certain cases for reasons, medical, e.g. risk of deformity, etc., or non-medical— e.g. where the marriage of the couple breaks down or the wife conceives after entering into the contract and so on.

The question then arises of how far such contracts are legal? Suppose a surrogate refuses to hand over the child on birth, can she be compelled to do so? The couple could also refuse to accept the child. And what happens if the child suffers from an infirmity and none of the parties wishes to own him?

Being the biological mother of the child the surrogate cannot be compelled to give away the child. So far as consideration given or agreed upon is concerned it won't be enforced as being against public policy. Children cannot be sold or bought. In Kentucky, for instance surrogate motherhood contracts have been declared illegal and unenforceable by the Attorney General. So if the couple refuses to take the child, the surrogate may keep it after being adequately reimbursed. And if she also is not interested in keeping the child it may be placed for adoption.

Then again there is the issue of paternity of the child. Under the law of evidence a child born within wedlock is presumed to be the legitimate child of the couple. In this case, suppose a married woman agrees to act as a surrogate (Normally married women are preferred because they have proven themselves to be physically capable of conception and child bearing; besides, those with children of their own are less likely to refuse to hand over the child on birth.) then, the presumption is that her husband is the natural father of the child. It remains for the genetic father who had engaged the surrogate to rebut this and prove his paternity. In any case the fact that the surrogate is the biological mother needs no testimony. So, in case of dispute, the question might arise between the biological father and the mother (surrogate) as to who is better suited to take custody of the child?

Another aspect of surrogate motherhood which needs thought is the risk of trading and commercialisation. If surrogate motherhood becomes a business the quality of children is likely to deteriorate. A surrogate may not be truthful about her medical history and also may be less careful about her diet and personal habits during the pregnancy.

It is also a well known fact that adopted children often suffer an identity crisis when they learn that they are adopted. It is a traumatic revelation for the child. In the case of a surrogate child, if the child comes to know of the fact, he may regard the surrogate mother as his true mother even though she is only the gestational mother. He might even suffer a sense of abandonment by the surrogate which might cause psychological problems. Similarly, the surrogate may also find it difficult to relinquish her ties with the child.

Another pertinent question in this matter is in regard to consent. Suppose the surrogate agrees to do the job without her husband's consent? Similarly, suppose the father enters into the contract with the surrogate without his wife's consent? In the former case, could the woman be charged with adultery? If she is compelled to abort under pressure from her husband, would she be liable to pay any damages to the couple who hired her services, and so on? And in the latter situation, suppose the wife refuses to accept the child? Issues of such nature can be numerous.

Yet another problem might arise. Suppose the surrogate dies during childbirth or suffers from some serious or permanent injury or disability, would the couple be liable? In case of her death, would her heirs be entitled to claim damages or compensation?

A day may come soon when our courts will have to resolve such issues. The problems mentioned above are just a few amongst the innumerable that might be linked with the subject, so it is high time some thought be given to them.

CHAPTER 3

MANUFACTURED CHILDREN: LEGITIMACY AND RIGHTS*

Procreation is one of the main purposes of marriage. When this end cannot be achieved owing to any physical or medical impediment, the childless couple resort to adoption. The adopted child is treated as the natural child of the adoptive parents for all practical, legal and worldly purposes.

The practice of adoption, however, is being fast superseded by other methods of substituted parenting. The reasons for this could be various — egoistic, dearth of children available for adoption, expediency or psychological.

The earliest technique of 'manufacturing' children is artificial insemination (AI). A couple who is not able to achieve conception through the normal natural course resort to this. If the husband is not normal, then the seed of a third person is injected into the wife to cause conception; if there is no medical problem with the husband then his own sperm may be used. Artificial insemination may be resorted to even when there is some medical impediment in the wife. This device of therapeutic impregnation by insemination can be successful in such cases too. Sometimes there may be no abnormality with either of them but due to psychological reasons, they fail to have a child; in such a situation also artificial insemination may be successfully tried.

* *Hindustan Times*, January 11, 1985

Convenient Mode

The next device is surrogate mothering where a third woman is hired and artificially impregnated by the husband's sperm and she carries through the pregnancy. Upon delivery, under a contract, the child is given to the couple who adopt the child. This may be done when the wife is unable to conceive. It is interesting to note that this device no longer remains an arrangement by necessity but also one of convenience. A wife who is career-conscious or working or in fear of labour etc., may find it a convenient mode of mothering a child.

In fact, surrogate mothering is a counterpart of artificial insemination. In the former the services of a woman are used by a couple, whereas in the latter, it is the male who serves the couple by donating his seed. It may be pointed out that there is at least one definite advantage of these two techniques—the child is at least 50 per cent the child of the couple. In artificial insemination the mother is the biological parent and in surrogate motherhood, the father is biological father.

In vitro fertilisation (test-tube babies) involves fertilisation of a male sperm and female ova outside the womb and the same is carried through the period till delivery in the womb of the wife or if she is incapable, in the womb of another woman. Similarly the process can involve the sperm of the husband, the ova of a third woman which is fertilised, and the womb of the wife who carries the child through the period of gestation and delivers. Thus the child may be the combination of both the parents, one of the parents or even none of the two (where the husband is impotent and even the sperm is of a third person).

Thus, these devices are purely mechanical and technical. Life is initiated as a result of a mechanical process divorced from the spiritual, mental, emotional and physical lives of the parents whose child the product of the device becomes. Consequently issues of diverse nature are bound to arise with them.

Crucial Issue

A very crucial issue that arises in such cases is in regard to the status and welfare of the children who are so 'manufactured'. They are not an outcome of any labour of love but a 'production'. Questions of their rights, interests and welfare may be ignored because the persons who are expected to care are the parents and in such cases there are in-built conflicts of interests and problems. To take a simple example — a surrogate carries a pregnancy to term and delivers. The couple who hire her back out and refuse to take the child for whatever reasons — may be the wife has conceived or the child is not up to the mark. Where does the child go? The surrogate may be an unmarried girl and may not want to keep the child. Even if she is a married woman, she may not want to be burdened with the child. Obviously the child is emotionally abandoned. It might be sent to an orphanage or some other institution. There is hardly anybody to take up the matter of such a child to the court — all persons who are related or were expected to be interested in him, have in fact only a superficial or mechanical tie and their interests may not coincide with the interests of the child. Apart from this problem, children born out of these therapeutic techniques become a subject of various other legal controversies.

The most important issue is in regard to the legitimacy of such children. In AI the child is born to a couple with the sperm of an outsider. Can the child be considered to be child of the marriage even though he is not the husband's child? Would it be a legitimate or an illegitimate child? If the donor is a married man, would not the child be considered to be the child of that couple? Applying the principles of the law of evidence, any child born during wedlock would be treated as the legitimate child of the couple (though this is rebuttable by proof of certain facts). However, in such cases biologically speaking the husband has no role in the conception. Would this child be considered to be a child of an adulterous relationship? But then, there is no consensual intercourse between the wife and the sperm donor to label the relationship as adulterous. These are some of the enigmatic situations.

In some cases (Canadian and American) the courts had occasion to determine the question of legitimacy of such children and divergent versions were given. In *Doornbas* v. *Doornbas* (1964), *Gursky* v. *Gursky* (1963) and *Anonymous* v. *Anonymous* (1954) the American courts held that such children are not legitimate since the father is not their biological father. In some later cases, however, the court's approach was more logical and where therapeutic impregnation was resorted to by the consent of the husband, the child so conceived was considered to be the legitimate child of the parents (For example, *People* v. *Sorensen*, 1968).

Children's Custody

Similar issues are bound to arise in surrogate mothering too. Who is the mother of the child, the one who carries the child or the one who takes over and brings up? So also in *in vitro fertilisation* where the egg and sperm may be fertilised mechanically and carried to term in the womb of another woman, problems of maternity are bound to arise. If the ova used is of one woman and after fertilisation it is implanted into the body of the other, who is the mother, the genetic or the gestational?

Question of custody of children arises in case of conflict between the parties. In some cases it has been held that since the child is not legitimate the father has no right of custody or visitation (*Doornbas* case). In others it was held that when such device was resorted to with the consent of the husband, he has rights over the child. In one case (*Strnad* v. *Strnad*) the New York Supreme Court, as early as 1948, observed:

> "The child has been potentially adopted or semi-adopted by the defendant (father). In any event . . . and with particular reference to visitation, he is entitled to the same rights as those acquired by a foster-parent who has formally adopted a child, if not the same rights as those to which a natural parent . . . would be entitled."

Dominant Criterion

In one case the woman who was hired refused to hand over the child to the couple who had 'engaged' her and in a custody claim, she succeeded against the couple. What would be the position vis-a-vis the husband and wife in case of divorce or dispute, and between the 'mother' and the 'surrogate' if the husband dies and both claim the child? In the former it is conceded that the position of the 'mother' should be the same as of any mother in normal custody disputes. A child may be as intimately and closely attached to the psychological mother as to a biological mother; it all depends on the degree and extent of the emotional relationship. So the claim of the 'mother' against the father cannot be pushed back only on the ground that the child was born through a surrogate and not the wife (this 'mother').

The same principle should apply in the other situation where the dispute is between the 'mother' and the surrogate. In any case, the dominant criterion is the welfare of the child and not the legal claims of the respective parties. So far as maintenance is concerned, the consent of a husband to the wife's artificial insemination is considered to be an implied promise and undertaking to support and maintain the child born to the wife. In surrogate children, the father in any case is the biological father and thus bound to maintain the child.

Another issue in this context relates to the inheritance rights of such children. Obviously a child conceived through artificial insemination cannot be expected to have any claims to the properties of the sperm donor. Moreover, the donor is generally not known. There is no reason why the child should be deprived of these rights. The mother is the biological mother and so there is no problem. The father, if he consents to the insemination, is deemed to be the father of the child for all practical and legal purposes and so the child's right to inheritance should not be in doubt. Even under the family laws normally illegitimate children, by which is meant children born to a couple out of wedlock or out of a void marriage, are entitled to inherit from their parents. There is another situation however where the position may be different viz. where a wife resorts to AI without the husband's consent or against the wishes of the husband. It is logical that in such situation the child's claim to properties of the father should not be pressed. Also, the child would have no rights to the properties of the surrogate mother's husband.

A very important emotional aspect of such children is the repercussions should the child come to know the circumstances and process leading to their birth. It is a traumatic revelation for the child to know that his father is not his biological father or the mother is not his biological mother. In case of the mother he may even suffer a sense of abandonment. All this creates psychological problems.

Law, legislation and judicial reasoning should keep pace with the fast medico-scientific developments and not lag behind. There should be proper safeguards against commercialisation and baby-sale and also for assuring the health and welfare of children so conceived and born. There should be clarity on questions of their maternity, paternity and legitimacy. Such children are already a product of mechanical or artificial arrangement, it is disconcerting that they should become an object of dispute too. It is the prime duty of the State and court to see what is in the best interests of the child ignoring all the claims of the selfish parties who "requisition" or "manufacture" him and then fight over him and drag him into a court.

CHAPTER 4

WHOSE CHILD IS IT ANYWAY?*

Advertisements play a very important role in boosting the popularity and sale of products these days. In fact, ads can be quite exciting at times, specially those selling a unique product, such as these: "Wanted to bear child for infertile couple; fees paid . . ."; "Surrogate mothers needed for infertile couples. Artificial insemination; must be healthy... medical expenses and living expenses ($800 per month) paid for 10 months". Sounds rather curious but the practice of hiring wombs, and producing children is very much in vogue and becoming increasingly popular in some countries.

Whereas certain states in the US have regulated the practice of surrogate mothering by statute, in Britain it has been given judicial recognition only very recently. Surrogate mothering is a process whereby a woman is hired by a couple, or even by a single person, to bear a child for the hirer. The woman is artificially impregnated with the sperm of the natural father; she carries the foetus to term and agrees to surrender the child upon birth. Although the practice sounds rather strange and novel, it has already gained popularity and according to some sources at least 2,50,000 Americans now living are a product of this process.

One of the first recorded instances of surrogate motherhood is found in the Bible. Abraham's wife, Sarah, who could

---

* *Indian Express*, February 17, 1985

not conceive, sent her husband to the Egyptian maid, Hagar, saying, "It may be that I may obtain children by her," and Hagar bore Ishmael. This is almost 4,000 years ago. The difference between that process and the present is that Ishmael was conceived by a natural process whereas the process of surrogate mothering at present is by way of artificial insemination. It is interesting to note that in Topeka, Kansas (US), there is an institute known as the Hagar Institute, which derives its name from the Biblical event and provides a surrogate mother service.

Quite often, apart from medical reasons this practice is resorted to for convenience, when a woman is working and does not want pregnancy to hamper her career or even when she is afraid of labour.

There is a considerable amount of debate on the propriety, morality and legality of surrogate motherhood. The protagonists feel that it is a great boon to the childless couple and if all concerned parties enter into a contract and abide by its terms there should be no objection from any quarters. Unlike in the case of adoption, where the adoptive parents and the adopted child have no biological relation, in a surrogate contract, at least the father is biologically related to the child. The surrogate mother hands over the child to the couple who hired her; the father is the natural father of the child whereas the mother's status is that of an adoptive mother.

The antagonists of this procedure have raised several questions. They fear that this practice will lead to the commercialisation and sale of babies. This objection is quite valid. Unscrupulous people might make it a business and womens' bodies may become an object of rent or lease for 10 months, just like houses. Babies may start being sold in the market like other commodities. As in the case of other products, the price demanded as rental and for sale of baby may vary depending on the "quality".

Then, there is the problem of enforcement of the contract. Suppose the surrogate mother refuses to deliver the child on birth? There have been a few such cases. In one case a surrogate, who was about to deliver a baby to a New York couple, changed her mind and decided to keep the baby. In *Syskowski* vs. *Appleyard* (Michigan, 1983) the court refused to give custody to the father against the surrogate mother who did not want to part with the child. Similarly, in England, a prostitute hired by a childless couple, who had agreed to bear and deliver the child, refused to part with the male child she gave birth to. The couple went to the court but lost.

Last month, a High Court in England which had taken temporary custody of a five-day old baby girl, born to a woman paid to a surrogate mother, ruled that the child be handed over to the natural father and his childless wife, both Americans. The judgement was delivered despite widespread "public hostility on commercial surrogacy on moral and legal grounds". In this case the couple had reportedly paid £14,000 to the surrogate parenting centre which includes the fee and medical and legal expenses. Thus, the British court has put the seal of legality on surrogate mothering.

However, a situation may arise, and a very unfortunate one, where the child, for some reason or the other, is not wanted anymore by the commissioning parents and nor does the surrogate want to take up its burden. This may happen if during the period of the surrogate mother's pregnancy, a child is conceived by the hiring couple or if the child is born with some deformity or medical problem. What is the fate of such a child? The question is, can any of the parties be compelled to keep or take over the child?

Then, there might be problems about the child's legitimacy. So far as the father is concerned, it is fine, since the sperm donor is the biological father, but what about the mother? Since the child is born out of wedlock so far as the commissioning parents are concerned — the child not being the child of the sperm donor's legally wedded wife — would it be considered to be illegitimate? Viewing it from the other angle — if the surrogate is a married woman and she had undergone the process with the husband's consent, then strictly speaking the child is their legitimate child in so far as it is born within wedlock. If the sperm donor backs out and breaches the contract, the husband of the surrogate gets into difficulties he had not envisaged earlier. He is legally responsible for the support, maintenance and custody of the child.

Another question that remains vague in law is in regard to custody claims over the child — the biological father vis-a-vis his wife in case of divorce or annulment and between the biological mother and the "mother" (the father's wife) in case of the father's death.

It is evident that as medical science and awareness of the technique advances, more and more childless couples will go in for surrogate motherhood. Even if the law tries to ban it, the practice will continue underhand. So, what cannot be banned *de facto* should be regulated *de jure* in the interest of the parties to the contract and more so in the interest of the "product" of the contract — the child.

In the face of legal vagueness an obscurity, the child suffers the pull of custody litigation, abandonment, rejection, and many other unhappy situations. What is needed is certainty, codification and proper regulation of the practice which has more than gained ground. Statutory regulations in some American states and the seal of legality granted to it by a British High Court on January 14, 1985, should be enough incentive to pave the way for similar legislation in other countries as well.

CHAPTER 5

WILL THE REAL MOTHER PLEASE STAND UP !*

Writing on surrogate mothering, prophesying a socio-medico-legal issue that would become a reality in the context of our society very soon and pointing out its legal implications, I was labelled as a writer who lives in a world of imagination, digging up issues which are too pre-mature and absolutely unrelated, for at least a few decades to come, in our society. And imagine, a bold advertisement appearing in a popular English daily a few days back which said:

"Well situated couple without children seeks respectable woman under 30, with at least one child, to act as surrogate mother. Woman selected will undergo artificial insemination conducted in clinic by woman medical doctor. There will be no contact with male donor. Application must include your date of birth, height and weight, education and working and background, your children's ages . . . . This is a sincere and genuine offer from a respected family. One lakh rupees fee . . . ."

Surrogate mothering as an alternative to natural biological mothering is fast gaining popularity in some countries. There are statutory provisions in the United States and the practice has been judicially recognised in the UK too. In the United States, a company known as "Surrogate Mothering, Ltd." was

* *Hindustan Times*, June 21, 1987

formed in 1981. A sensational news item in a popular daily said:

> "A new company has been formed and expects to deliver 12 of its products — babies — soon. An additional 20 have been ordered."

Now that a beginning has been made in this direction in our country as well it is time that legal implications of the process be kept in mind.

Suppose a woman offers herself as a surrogate and agrees to undergo the process of artificial insemination. The process might have to be repeated a couple of times; how long should the couple wait before the surrogate conceives? Say after six month's or a year if there is no headway and the parents cancel the contract what are the rights or claims of the parties. Can the surrogate ask for any fee for undergoing that process?

If the woman conceives but has an abortion, is she entitled to any damages; or conversely if after conception she herself goes in for an abortion, do the couple have any cause of action against her? Suppose the woman suffers some disabilities during pregnancy or dies during that period, what is the legal position? What if the surrogate mother refuses to hand over the child after birth? Can the parents file a suit for specific performance. In fact there have been some cases of this kind in America and England.

In one case, a surrogate who was about to deliver a baby to a New York couple changed her mind and decided to keep the baby. In *Syskowski* v. *Appleyard* (Michigan, 1983) there was a dispute between the surrogate mother and the father over the custody of the child. The court refused to give custody to the father.

In England, a prostitute was hired by a childless couple who had agreed to bear and deliver the child. After birth of the child, she refused to part with it. The couple went to the court but lost. In England recently, the High Court ruled that the surrogate mother should hand over the child to the couple who had hired her. The couple had reportedly paid £14,000 to the surrogate parenting centre which included the fee and medical and legal expenses.

A very pertinent issue or rather an unfortunate situation may arise where a couple, who hire a surrogate, refuse to take a child. This is very likely in a situation where a child so born does not belong to the gender of choice, or suffers from a medical problem, or where meanwhile the couple have their own child. Where does the child go when both parties disown it or refuse to own it? In this case the surrogate has performed her part of the contract but the couple back out by not accepting the child or not paying the amount agreed upon.

Problems of paternity of the child may crop up too. A child born within wedlock is presumed to be the child of the bearer's husband under the law of evidence. Suppose he refuses to part with the child, what are the claims of the biological father i.e., the husband in the hiring couple?

These and innumerable other issues are likely to come up in the transaction thereby giving rise to judicial resort — which undoubtedly is a very unhappy state of affairs.

It is true that surrogate mothering is one of the most satisfying procedures filling the void in the life of a childless couple. In adoption, the child is a complete outsider whereas in a surrogate, the child is 50 per cent the biological child of the couple as the husband is the father of the child. Besides, getting a child in adoption is not very easy since there is a long wait for this. Thus a surrogate child has an edge over an adopted child.

What is needed is a proper regulation of the prcactice. The rights and liabilities of both the parties should be clearly laid down. The written consent of both the spouses in both the parties i.e. the husband of the hired woman and the wife of the hiring couple — should be mandatory. Besides only those couples who have been medically declared incapable of having children should be permitted to hire a surrogate. The entire process must be performed in strict secrecy and confidentiality. The child should not know about it. Strict guard is also needed against exploitation, trading and commercialisation of this practice. When individual cases have started coming up, the day is not far when firms will be set up for this purpose soliciting clients — both childless couples and surrogate mothers. And once the process starts, legal implications are bound to arise.

CHAPTER 6

ARTIFICIAL INSEMINATION ACT: 1995*

Advancements in medical science have given hope of progeny to many childless couples who are either infertile or have problems in natural normal procreation or conception. The use of assisted reproductive techniques is a great boon for such couples.

Artificial insemination (AI) is one of the common methods used by infertile couples. The history of AI dates back to 1322 when mares were successfully bred by Arabs. The superior mares of the enemies were artificially impregnated by the sperms of inferior horses to weaken them. From this time onwards until the middle of the sixteenth century, successful experiments were carried out on fish and animals.

The first successful AI of a human being is credited to Dr John Hunter, an English physician, who used this device on a married woman using her husband's semen in the late 18th century. In 1890, Dr. Robert L. Dickinson established the practice of AI using donor's semen. However, it was not until the 20th century that AI for the purpose of procreation was perfected.

AI can be practised in three ways: AI using husband's semen (AIH), AI using third party donor's semen (AID), and AI using mixed semen of husband and third party known as combination artificial insemination (CAI).

---

* *Hindustan Times*, October 18, 1995

While the first two are common, CAI is not; and again, even between the AIH and AID, the former has been more acceptable than the latter because of obvious reasons.

Though this medical technique has been commonly used in other countries, its use in India is restricted.

Many childless couples with resources have no doubt been seeking aid of assisted reproductive techniques but, in the absence of clear guidelines and explicit legal sanction, the use has been more surreptitious and subdued involving the risk of exploitation, commercialisation and medical hazards. Realising all this, the Delhi Legislative Assembly has enacted the Delhi Artificial Insemination (Human) Act 1995. The salient features of this Act are:

It legalises donation of semen and of ovum. Banks for storage, sale, donation and supply of semen have been recognised with provisions for their compulsory registration.

There is to be a Supervisory Authority, who is the Director of Health Services. In case of violation of any provisions of the Act, the registration can be cancelled. There is also a provision for punishment with imprisonment for a term which may extend to three years and a minimum fine of Rs. 5,000.

Not unoblivious of the disastrous medical and social consequences which could follow the irresponsible acceptance of semen donation and indiscreet performance of artificial insemination, the Act provides for strict testing of the semen against HIV infection and also prohibits the segregation of the XX or XY chromosomes so that pre-conception sex selection is not possible. (XX chromosomes denote female sex and XY chromosomes denote male sex.)

In our society where people suffer from the son-syndrome, preconception male sex selection and female sex rejection is very likely if segregation is not banned.

The Act also provides for maintenance of secrecy about the identity of the donor and the recipient. At the same time however, it also provides that "complete record of the biodata including mark of identification of the donor and recipient of the semen or ovum" shall be kept by the semen bank, doctor or hospital performing AI. One wonders how confidentiality can really be maintained with the various escape routes in the set up. Strict rules have to be adhered to if the provision of confidentiality is to be made meaningful.

The enactment of the law is not the end of all problems but also a beginning of many. Though it provides that the written consent of the husband and wife seeking AI should be taken, yet, there are various legal, ethical, social and moral issues which can arise out of the practice. For instance, suppose a doctor performs AI without the consent of the husband, would it amount to adultery on the part of the wife? (The doctor is liable to punishment in such a case.) What would be the status of such a child? Also, would AI amounts to consummation of a marriage and debar a wife from seeking matrimonial relief against her husband on ground of his impotency or non-consummation? Can the wife resort to AIH after the death of her husband? This is now possible since the semen can be stored and preserved in the semen bank. Would such child be posthumous child? What is the extent of liability of the donor or doctor performing AI of infected semen resulting in the birth of a deformed/diseased child? Can the parents or the child sue the donor or doctor for the negligence? These and many more issues will arise which can be settled only in due course as and when they come up for adjudication or clarification.

Another aspect of artificial reproductive techniques, in our context especially, is whether with our demographic profile — population explosion and economic conditions — our country needs, or rather can afford, such techniques. Our children are languishing in the streets, begging, undernourished, and unclad. Would it not be better to encourage adoption of those already born and give them life than add to the population? But perhaps this is utopian sentimentalism. Having a child of one's own brings more joy and satisfaction than adopting and bringing up as one's own child who is born to someone else. The Act may satisfy such need of many childless couples but only its working over a period of time will prove or disprove its expediency in our country.

CHAPTER 7

SURROGATE MOTHERHOOD: IMPLICATIONS AND ISSUES*

Introduction

Medical science and technology have made remarkable progress during the past few decades. The once unimaginable is a reality today with all its ramifications and implications. The pace of these developments being at times faster than societal thinking and legal changes, they are bound to give rise to social, legal, moral and ethical dilemas.

An important area of medical research and advancement which has changed the lives of innumerable infertile couples is assisted reproduction.

The desire for progeny being a normal natural desire of almost every human being, the inability or disability to found a family could indeed be devastating, emotionally as well as socially. In earlier times of comparative ignorance of the cause and cure for infertility, childless couples had no option but to resign to their fate, or occassionally, resort to adoption. But things have changed now. Science has entered God's domain. The creation of a new life which was entirely God's grace can now be the result of combination and permutation of several medical techniques and technologies without actual physical

* Paper presented at the National Seminar on Bioethics: Indian Perspectives, November 26–28 1997, Goa.

union of the male and female. In an age of synthetics we find ourselves heading towards synthetic babies too.

This short paper, however, attempts to deal with a mode of reproduction which could be a combination of artificial and natural processes of conception and procreation. This is surrogate mothering.

Surrogate Mothering

Surrogate means substitute. Thus a surrogate is a substitute for the wife to carry a child. She is a woman whose womb is hired, and she is inseminated with the sperm of a man, not her husband, to carry a pregnancy and deliver the child to that man and his female partner under a pre-conception adoption agreement.

The practice of surrogate motherhood has historical precedents. The sixteenth chapter of Genesis tells the following story about Abraham and his wife Sarah.[1]

> "Abram's wife Sarai had borne him no children. Now, she had an Egyptian slave girl whose name was Hagar and she said to Abram, 'You see that the Lord has not allowed me to bear a child. Take my slave girl, perhaps I shall found a family through her'. Abram agreed to what his wife said, so Sarai, Abram's wife brought her slave-girl, Hagar, the Egyptian, and gave her to her husband Abram as a wife . . . and she conceived."

In earlier times, the process of conception used to be by natural intercourse by the husband of the infertile woman with the surrogate. However, medical developments today have dispensed with the necessity of a physical relationship. A surrogate now normally agrees to be artificially inseminated and in accordance with a pre-conception agreement, hands over the resultant child to the genetic father and his wife.

Surrogacy may be either full or partial. A full surrogacy is one where the surrogate donates her gamete and also carries the embryo which has been fertilised with the man's sperm. Thus, she is the gestational as well as the genetic mother. In partial surrogacy, however, the surrogate lends only her womb. It is pertinent to note that the technique of *in vitro fertilisation* (IVF) of human oocyte with human sperm was initially performed with the intention of helping couples who suffered handicap to procreate naturally. Thus the fertilised pre-embryo could be placed into the uterus of the woman who provided the gamete. The existence of a pre-embryo outside the body of the woman, however, creates the possibility for the pre-embryo to be gestated by any other woman and then returned to the genetic mother after birth. Thus arose the concept of partial surrogacy where the woman is only the gestational but not the genetic mother of the child. The first reported pregnancy of a surrogate gestational mother was in 1985, which culminated in successful birth in April 1986.[2]

Hiring a womb to have a baby is seen as the best option for a couple to have a child almost immediately after birth and that too with a genetic link with at least the father if not the mother too (partial surrogacy). Prior to this medical advancement, adoption was the only way out. Adoption however entailed a long waiting period. Moreover, it could not be emotionally very satisfying since the adopted child has no genetic link with the parents. In any case, now, in view of the use of contraceptive devices, availability of medical termination of pregnancy procedures and changed attitudes towards single mothers, the number of children placed for adoption is not very large. No wonder, surrogacy arrangements are becoming more popular.

It cannot however be denied that surrogate mothering is a challenge to the traditional concept of motherhood and family and the same is bound to give rise to queer situations and complex issues.

Implications

The impact and implications of surrogacy arrangements need to be viewed from the perspective of the commissioning couple, the surrogate, and the child so born, as also the social, legal and ethical issues involved therein.

Commissioning Couple

Viewed from the commissioning couple's perspective, surrogacy arrangement is a very effective means of enabling a childless couple to have a child. Prior to the development of medical technologies, adoption was the only way a childless couple could have a child and bring it up as own. Adoption however, as pointed out earlier, is neither very easy nor satisfying. Rearing an adopted child is not the same as having and rearing one with a genetic link. In full surrogacy, the father is the genetic father whereas in partial surrogacy, both the mother and father are the genetic parents of the child gestated by the surrogate.

It cannot however be denied that such arrangement also has a potential of harm to the couple who may live under constant tension and suspense till the child is handed over. They may be harassed or blackmailed by the surrogate; the surrogate might refuse to relinquish the child; or even after handing over the child, she may continue contact with the couple or child thereby causing deep embarassment and emotional disturbances in the family. As remarked by Justice Latey in *Re Adoption Application* (Surrogacy):[3]

> "... before they go down the path of surrogacy they should know and know fully what it may entail. It is not a primrose path."

Another problem that may come up for the couple is that if the surrogate refuses to give up the child the contracting father may still have to pay child support because he is the biological father. Though under the terms of agreement he may have been completely absolved of any parental obligation, yet harassment cannot be ruled out.

A reference may be made to some cases which are indicative of the stress and emotional trauma that the commissioning couple might be subjected to in the process. Noel Keane, an American lawyer who arranged surrogacies wrote a book *The Surrogate Mother* in which he shares some experiences.[4]

> 'Bill and Bridget were an infertile couple. Keane put them in touch with Diane. Diane had seen a television talk show in which Keane appeared with another infertile

couple and their pregnant partial surrogate. After the show, Diane had phoned offering to help a couple to have a child. She was a thirty one year old divorcee . . . with a two year old son: when interviewed she seemed a good mother and responsible parent. At that time Keane had not discovered that Kentucky State law allowed the payment of a fee to surrogate mothers, so Diane was told that nothing could be paid except expenses. She signed an agreement to that effect.

Diane soon became pregnant with Bill's child. Then things began to go wrong. She asked Bill and Bridget for money to travel to Boston to visit her mother. They sent her the money. She said she had been robbed of expense money they had sent her. They sent her another cheque. Then her car needed repairs, she had extra medical expenses . . . and so on. Often when she phoned asking for money, she sounded drunk or stoned on drugs. Sometimes she threatened to kill herself unless she got more money. Bill and Bridget did not dare call her bluff. Shortly before the baby was due, Diane demanded $ 3,000 to pay for a computer course she planned to take. Bill and Bridget paid. In all, they calculated that they sent Diane more than $ 12,000.

Finally, two weeks before the due date, Diane phoned to say that she was in jail on a drunk driving charge, and needed bail money. Bill and Bridget flew to Tennessee to stay with Diane and try to prevent anything else going wrong before she had the baby. That is when they found that Diane's 'roommate', Vicky, was really her lover. In despair Bridget turned on her husband. 'Bill', she screamed, 'do you realise that the woman you got pregnant is not only an alcoholic and drug addict but also a lesbian!'

Amazingly, the story has a happy ending. Diane gave birth to a boy, below normal weight and suffering from withdrawal symptoms, but after five days in hospital Bill Jr. was healthy enough to go home with his father and his new mother. Diane tried for some time to extract more money by threatening to hold up the adoption proceeding but when this threat failed to have any effect, she moved interstate without leaving any forwarding address.

There are also high chances of a court battle over child's custody and wardship. To take some examples : In *A* v. *C*,[5] the surrogate changed her mind and refused to hand over the child. On father's application, access was granted to him in the child's best interest but the court of appeal unanimously reversed the decision. Omrod L.J. observed:[6]

> "I can see absolutely no advantage to this child in continuing to be in contact with the father, except possibly a financial advantage to which I attach no significance whatever, in this case. If the father is to continue to turn up in the mother's house or to keep meeting her somewhere to take over this child, or to meet some member of her family to take over the child and return the child, the whole of this sordid story will be revived weekly or monthly as the case may be. The mother's position will be handicapped, and the handicapping of her position handicaps the child."

Again, in *Re P*[7] (minors) a woman offered her services as a surrogate mother to a married professional man who donated sperm by artificial insemination. During the pregnancy, she began to have misgivings about giving up the child. However, once she delivered (she had twins) her reluctance to part with the babies became stronger. Though she regretted disappointing the father, she decided to keep the children. The court after weighing all the factors ordered the care and control of the babies to their mother "to whom they are bonded".

In *Baby M*,[8] after a rigorous court battle, custody of a female baby born to a surrogate by arrangement, was ordered to be given to the commissioning couple. While acknowledging that surrogacy contract itself is unenforceable and illegal, yet because the welfare of the child so demanded, custody was awarded to the hiring couple.

In *Calvert* v. *Johnson*[9] a couple hired a gestational surrogate to carry through a pregnancy with the gametes of the hiring couple. In other words, the hiring couple were the genetic father and mother whereas the surrogate was the gestational mother of the child. The surrogate, contrary to the agreement, wanted to keep the child. She fought for the custody of the child. After a long battle, the California Supreme Court ruled against the surrogate and upheld custody award in favour of the commissioning couple.

Surrogate Mother

A. Advantages

Income source

An argument in support of surrogacy from the surrogate's perspective is that it adds to the income options of a woman who may be in need of money. Some divorced women with young children have, reportedly, chosen to be surrogates in order to support their children and look after them while remaining at home. In 1981, Judy Stiver,[10] a Michigan housewife responded to an advertisement seeking a woman to become a surrogate because it "seemed to be a good way to earn some extra cash." In her own words "we wanted the money to pay some bills and take a vacation." Womb lending for financial reasons is not unknown even in India as indicated by the much publicised case of Nirmala from Chandigarh. In fact monetary incentive is most likely to be the reason behind rendering of these services in our country because of the economic conditions here.

Nirmala, reportedly[11] agreed to act as a surrogate for an NRI couple in return for Rs. 50,000 because she had a child and an ailing husband for whose treatment she was in dire need of money. The whole plan however collapsed probably because she agreed to conceive by a natural act and sought advise on the legality of the whole procedure.

Altruism

A woman who offers gestational services to a childless couple by carrying a child having a genetic link with them might derive a satisfaction of having given happiness to them. There is an altruism involved in the gesture. As observed by a surrogate mother who carried a child for a couple in *Re Adoption Application (Surrogacy)*:[12]

> I wanted to help a childless couple. My own children are very precious to me and I sympathised greatly with any couple who were unable to have children of their own, so much so that I was willing to have another pregnancy in order to give someone else that joy.

According to the report of child's guardian ad litem:

"Her interest in surrogacy [is attributed to] a particular pleasure she has in having babies and a great sympathy for women who are unable to experience the joy of having and caring for a baby."

Psychological satisfaction

One preliminary study[13] in America found that one third of surrogate mothers may be using the process to help themselves psychologically. There are women who in the past have voluntarily aborted or given up child through adoption and have then become surrogate mothers in order to relive the experience of pregnancy in a psychologically satisfactory way.

Whatever the arguments in support of surrogacy from the surrogates perspective, it cannot be denied that it has also a potential of grave damage to her and her family.

B. Disadvantages

Emotional trauma of separation

The relationship of a surrogate mother with the child during and after pregnancy may be very ambiguous. She may start with feeling good about her gesture to help a childless couple but this may gradually develop conflict, guilt and apprehension in her mind. She may suffer great psychological or emotional harm by having to give up the child she has carried. Unlike sperm donation, which is a transitory act, or even an egg donation, which though more complicated, is not a continuing process, the use of a womb involves the womans' whole body. "Whether her pregnancy is achieved by artificial insemination or *in vitro fertilisation* with an egg other than her own, it is her body which nourishes the foetus, resounds to its heart beat and its first movements, and pushes the baby out into the world.[14] Thus, for a gestational mother to disassociate herself from the baby is neither natural nor easy. There have been instances where surrogates have gone through a period of grief and mourning after giving up the child and that is the reason for high chances of surrogates refusing to give up the child.

As observed by Justice Latey in *Re adoption [Surrogacy]*,[15]

> "In the nature of things, those who undertake the role of surrogate mother are those with strong maternal instincts. The likelihood must be there that when the baby arrives, the mother cannot bear to part with it. The trauma and turmoil . . . needs no describing."

Impact on family

Apart from this there is a likelihood of adverse consequences of surrogacy on the surrogate's family too. If the commissioning couple refuse to take the child, the surrogate's husband may be saddled with the responsibility of bringing up the child as the court cannot compel the hiring couple to accept the unwanted child. The child having been born to his wife during wedlock would be legally considered to be his legtimate child. In one case mentioned by Keane[16] in his *The Surrogate Mother*, a woman agreed to bear a child for a couple and abstain from sexual intercourse until the baby was conceived. In return, the father agreed to pay her ten thousand dollars and take the child. All went well until the baby was born when it was discovered that he suffered from a deformity. On the basis of blood reports, the commissioning 'father' claimed that it could not be his child and so refused to accept the child and also to pay the agreed sum. The surrogate couple also refused to keep the child as they already had a two year old daughter and did not want another child. When further blood tests were ordered by the court which confirmed that the child was not the commissioning father's, the surrogate had no choice but to keep the child.

Another effect of surrogacy on the surrogates family is that her children might feel neglected during her period of pregnancy. After the birth of the child the siblings might also suffer a dilemma at the sudden disappearance of the child born in the family.

The Child

Gestational risks

It is, however, from the child's perspective that the issue of surrogacy needs utmost reflection. In the entire procedure, all other parties have some interest or the other; the commissioning couple gets a child, the surrogate might derive psychological satisfaction or some economic gain and the doctor renders his professional services and earns. It is the child alone, the product of this arrangement, who has no say in the matter and is the one most affected too. Unlike as in adoption where it can be argued that a child gets a home and family, in a surrogacy arrangement there is no child in existence — one is specially created to be handed over to those who placed the order with all the hazards involved for the child.

Beginning from conception, the child is a product which has been agreed to be manufactured for some consideration. During the gestation period, the surrogate is only a mechanical carrier whose womb has been hired. There is a danger of inadvertent transmission of a genetic defect. The surrogate may be either unaware of the condition or she may even deliberately conceal the same for fear of losing the contract. Apart from this, she may be callous about her food habits and other activities which could harm the foetus and result in the birth of a handicapped child. Since the surrogate is only a mechanical carrier with no intention of rearing the child, the health and quality of such child may not be her concern. Those who support surrogacy, however, suggest that a thorough screening of the potential surrogate would eliminate genetic and gestational risks to the child. Likewise, under the terms of an agreement, restrictions can be imposed on her food and other habits and activities to ensure that no harm is caused to the foetus. Though these measures might, to some extent, diminish the hazard but obviously, constant monitoring of the woman is not feasible. This is clearly evident from Judy Stiver's[17] case where, inspite of an agreement to abstain from sexual intercourse until the child is conceived, it was discovered that the child born to her was not the child of the man who had given his sperms to father a child.

Ill Treatment

There is a risk of such child being ill treated[18] by the hiring couple more so if they have meanwhile begotten their own. In the case of the wife, since she may not have made any contribution, there is a likelihood of her not having much attachment for the baby. This argument however is repelled by a counter argument that such children are infact likely to be more cared and wanted. Parents who are unable to have their children in a normal natural way have gone through a stressful period, and spent huge amounts on treatment. They will, therefore, be more possessive and attached to the child and the child would also feel more wanted and important.

Adjustment problems

Surrogacy also invovles a risk of adverse effect on the child's psychology when it comes to know that it has a blurred geneology; he might be confused about his self identity. He is a product with at least two mothers, if not three, the gestational and the rearing and in some cases even a different genetic mother. The child may yearn to meet his birth parents; or he may suffer a feeling of abandonment and rejection by the birth mother while at the same time not able to emotionally assimilate in the family of the rearing parents. Such a child may have a split personality and suffer complexes and adjustment problems. Those who support surrogacy, however, argue that some of the risk caused by confused geneology may be outweighed by possible benefits of a child having parents who desperately needed him. A child conceived through this arrangement gets a healthier and more loving environment than one whose birth is unplanned.

The Issues

Apart from the implications of surrogacy on the parties directly involved, there are several other basic issues as well which are too significant to be ignored and too complex for an easy solution.

Commercialisation

The most important issue is the risk of commercialisation of human reproduction. The availability of contraception, abortion and the greater willingness of single mothers to bring up their children has led to a shortage of babies offered for adoption. That coupled with desperation of childless couples is paving the way for professional brokers to enter the field and facilitate surrogacy arrangements and agreements. Surrogacy is commercialisation of a service by hiring/renting a womb. The surrogate is the baby breeding machine; the couple are the consumers and the baby is the product which is sold.

Omrod L.J. described the surrogacy arrangement in *A* v. *C*[19] as "a sordid commercial bargain." Cumming L.J. described the arrangement as "a kind of baby farming operation of a wholly distasteful and lamentable kind" and a "lamentable commercial transaction".

Those who support surrogacy, however, argue that there is no commercialisation or sale of baby involved in this procedure as there is no child in existence when the contract is made. The woman only provides the gestational service to the man whose child she carries because his own wife is medically incapable of that. The monetary transaction that transpires is not for sale or purchase of a baby but to meet the expenses connected with the pregnancy. From the couples angle also, such payments are not considered to be unethical as, even without involvement of a surrogate, couples spend huge amounts on infertility treatments. Hiring a womb is seen as just one way to overcome the infertility problem.

Exploitation of Women

It is inconsistent with human dignity that a woman should use her womb for financial consideration and treat it as an incubator for some one else's child. Surrogacy degrades the personality of a woman. Elizabeth Kane, the first commercial surrogate in the United States, now active in the National Coalition Against Surrogacy, writes: "A woman (surrogate mother) feels like a flesh covered test tube during the entire experience. As the foetus grows, the woman is depersonalised, she becomes fragmented from the whole person — merely a vehicle for breeding babies."[20]

Organised racketeering in flesh and organ trade and exploitation of the economically weak by professional brokers and middlemen are proof enough of how human body and parts thereof could become objects of commerce. Poor women will be impelled to offer their gestational services to couples who can afford to pay. Even if there is a legal ban, poor women may be hired and exploited to enter into an agreement.

Thus, it is not only the morality of the procedure which is involved but also the possible exploitation of women, specially when payment is involved. And since surrogate is a woman, it could be viewed as gender based exploitation.

Population Perspective

From the demographic perspective one might wonder whether in a society which is already overpopulated, it would be desirable to specially create more human beings through such procedures. It is argued by some that national interest should have priority over selfish individual interests and so couples who are not able to have children in the normal natural course should not be permitted to resort to these procedures. It may however be argued that the number of children born this way is a very small fraction as compared to the naturally increasing population. So, why not put a restriction on family size rather than deny the joy of even a single child to couples who wish to resort to assisted reproduction?

Interference with Nature

Ethicists argue whether it is proper to interfere with nature and the will of God and resort to unnatural procedures for having a child. When the contraception and abortion issues were raised, the idea of sex without children was considered as unacceptable and now we have reached a stage when children without sex is a reality. The supporters however, do not find this argument of unnaturalness as tenable.

complicated infertility treatments are allowed, if the seriously ill can be put on expensive complicated medication and surgery; if the dying process of the terminally ill can be prolonged by the use of artificial respirators and machines, all of which are in a way an interference with nature, why then object to assisted reproduction? It is argued.

Ethics of Third Party Involvement

Then there is also the question about the ethics of third party involvement in procreation. Surrogacy requires the active cooperation of an outsider woman right from beginning till the handing over of the child. It is argued by some that this would destroy and weaken the bond of marriage and threaten its sanctity. Besides, it would undermine the integrity of the institution of marriage. Against this, however, it is felt that far from weakening the bond of marriage, the entry of a child in the family, albeit through a surrogate but biogically the genetic offspring of at least one of the partners, would strengthen family ties and add to its stability. It is well known that the absence of a child in a family could lead to marital discord and at times it might even lead to divorce. In such situation a child through a surrogate is definitely a better alternative than divorcing a wife and remarrying in order to have a child.

Non Medical Surrogacies

Hiring a surrogate by a couple for non medical reasons is another possibility. A surrogate may be used to produce a baby just for the convenience of the potential mother who does not wish to undergo the pain and inconvenience of a pregnancy. Such a woman is not likely to be a very competent and loving mother because she has neither suffered the pain or pangs of labour, nor the agony or devastating experience of infertility. While the use of surrogate even in the face of medical indications of disability is still a matter of debate, surrogacy for non medical reasons seems to be an absolutely unacceptable proposition. The U.K. Human Fertilisation and Embryology Act's *Code of Practice* (1993) clearly states that "The application of assisted conception techniques to initiate a surrogate pregnancy should only be considered where it is physically impossible or highly undesirable for medical reasons for the commissioning mother to carry the child." (para 3.19)

Parentage and Relationships

Assisted reproductive procedures have given rise to uncertainties as regards parentage of the child. A child born to a wife during wedlock is presumed to be the legitimate child of her husband even if there is no genetic link. Thus the biological father may be different from the legal father. This uncertainty has extended from paternity to maternity as well now.

A woman who bears and delivers is the biological mother but if the gametes have been provided by the hiring woman, she may claim to be the genetic mother of the child as happened in *Calvert v. Johnson*.[21] Relationships between the children born to a woman as a surrogate and as a mother is also a dilemmic issue. Surrogacy arrangements may sometimes lead to queer situations, e.g. in one case, a grandmother carried her own grand children for her daughter because the daughter was born without a uterus. The grandchildren were created from the gametes of the daughter and son-in-law.[22]

Legal Intervention

Whatever the issues or intricacies of surrogate motherhood, the fact is that it is seen as an effective solution to childlessness by those who can afford it. However in view of the complexities and issues that surround surrogacy arrangements, the intervention of law is inevitable. The question is, whether the practice should be totally banned or it should be regulated with adequate safeguards.

Complete Ban?

Though some countries have imposed a complete ban on surrogacy arrangements,[23] the viability and success of such measure is doubtful. Firstly, it may not work because a couple who is determined to have a child this way will easily arrange a surrogate for a price who will carry a child for them. The arrangement is too personal and private to be monitored. Secondly, under-ground surrogacy would make the detection of possible harms of the procedure almost impossible. Thirdly, if the practice is to be criminalised, there will be difficulties in deciding upon the sanction. Fine would be no deterrent for a couple who is prepared to spend huge amounts to get a child. A sentence of imprisonment for the mother would be against the child's interest who will be seen as an outcome of a criminal transaction. Besides, it will be extremely disastrous to insist that the child stays with a mother who does not want him, or declare him an illegitimate or refuse legal recognition as parents to the couple who claim the child.

Regulation

Since outlawing surrogacy does not appear to be a feasible proposition, the logical alternative could be a regulation of the practice. The objectionable components of the practice should be weeded out and the entire procedure could be regulated by law. Commercialisation being the strongest objection to surrogacy contracts, the same can be combated to some extent by putting a ban on intermediaries and agencies facilitating surrogacy contacts and contracts.

The Surrogacy Arrangements Act, 1985 was the first legislation on surrogacy in the U.K. The Act prohibits an agency or individual other than the potential surrogate mother or commissioning parents from acting on a commercial basis to initiate, negotiate or compile information towards the making of surrogate arrangement. The Act does not however, prohibit non-commercial agencies or the activites of anyone who might arrange a surrogacy non commercially nor does it ban payments to a surrogate mother.

Main purpose of the Act is to prohibit making of surrogacy arrangement on commercial basis. Advertising by way of newspapers, periodicals and telecommunications is proscribed, the criminal liability resting on the proprietor, editor, pub lisher etc. The principals involved are however expressly exempted from criminal liability [S.2(2)] Moreover, payments made to or for benefit of the surrogate are not regarded as being made on a commercial basis [S.2(3)]. The Act has however, confused the position of professionals such as doctors and lawyers whose services in any surrogacy arrangement are clearly desirable but who would be unable to be paid for intervention on behalf of any of the principals without contravening section 2 of the Act. Infact, if surrogacy is to be permitted, under whatever conditions, services of lawyers, doctors and counsellors cannot be dispensed with.

The Human Fertilisation and Embryology Act, 1990 also makes a surrogacy arrangement unenforceable in law (Section 36). Thus, so long as parties volitionally honour the terms of the agreement, there is no problem but if any of the parties at any point of time back out, the court cannot order specific performance of the contract.[24] The idea behind this provision is to discourage such arrangements. As aptly remarked by Justice Wilentz in the *Baby M.* case, "our declaration that this surrogacy contract is unenforceable and illegal is sufficient to deter similar agreements . . . ."

The Act also empowers the court to make an order providing for a child carried by a surrogate to be treated in law as the child of the commissioning couple provided that the gametes of one or both have been involved. This is subject to the consent of the surrogate and her husband. Such order will confer parental rights and obligations on the commissioning couple (Section 30).

Conclusion

That surrogate mothering is no longer an alien concept in India is evident from Nirmala's case referred to above as also from the ads for the service. An ad in a popular fortnightly reads:[25]

> "Wanted young woman 20–30 years
> who can bear child by artificial insemination
> for childless couple from North India.
> Confidentiality assured. Desirable award will
> be given. Box No. . . . . C/o. . . . . New Delhi.

Thus though not very openly, surrogate mothering arrangements do exist. There is however no legislation to regulate the practice. The Delhi Artificial Insemination (Human) Act, 1995 which provides for the regulation of donation or sale and supply of human semen and ovum for the purpose of artificial insemination makes no mention of womb — hiring or letting. Invoking general principles of law, however, it is obvious that no agreement which provides for the permanent separation of a new born from its mother to another, either for love or money, would be enforceable in a court of law. Commercialisation of the activity through agencies or intermediaries would be an offence and amount to illegal trafficking in children. Nonetheless, the very nature of the procedure being so personal and private, a complete legal ban would be an ineffective solution. We have the examples of pre-natal diagnostic procedures and organ trading. The Pre-natal Diagnostic Techniques (Regulation and Prevention) Act 1994 does not put a complete ban or pre-natal diagnostic procedures but only on the objectionable fall out of these procedures, viz. the sex determination tests. Likewise, it is not the supply and use of organs but its trading which is banned under the Transplantation of Human Organs Act, 1994. The same approach can be followed in respect of surrogacy arrangements too, like the United Kingdom. While private arrangements may be tolerated with proper regulation, trading and commercialisation should be banned and made a penal offence. Seeking to find a woman willing to perform such service for payment, or initiating or negotiating any arrangement involving the making of any payment for the service, or publication or distribution of an advertisement inviting women to offer these services should be made punishable. The interests of all parties — the hiring couple, the surrogate and most important the child — needs to be guarded. Children should not be irresponsibly produced. A woman who is to render gestational services should be thoroughly screened to eliminate risk of transmission of any defect to the child since an abnormal child is very likely to be rejected by the parties and we have no infrastructure to care for such babies.

Effective counselling to families of both the parties is very essential. The surrogate and her family must be apprised of the implications of the arrangement and the possibility of breach of the terms of the agreement by the other party.

Likewise the hiring couple needs to be counselled too. Screening of the sperm of the male and gestational capacity of the female must be strongly advised in the interest of the surrogate and the child to be born.

Intrafamilial surrogacy should be discouraged because it can cause more disturbances in the relationships and also open doors for emotional coercion, besides chances of incestuous relationships.

Once a child is handed over, the court should make a parental order in favour of the couple conferring all rights and liabilities in respect of the child on them and absolving the surrogate and her husband and family of any parental obligations towards it.

All said, the practice of surrogacy even with the best of safeguards has a potential for economic exploitation, moral confusion and psychological harm to the surrogate mother, the prospective adoptive parents and the children.[26] And with the pace with which the practice is becoming popular it seems the time may not be too far when we might find ourselves in a world where there will be a special class of women who will be child breeders for a price catering to the needs of couples who hire them for reasons medical or simply convenience. Scientific and medical progress must go on no doubt, but we should not be unoblivious of its perils.

REFERENCES

1. See Peter Singer and Deane Wells, *New Ways of Making Babies: The Reproduction Revolution* (1984) at 107–108. In Topeka, Kansas (U.S) there is an Institute known as the Hagar Institute which derives its name from the Biblical event and provides a surrogate mothering service.
2. *Fertility and Sterility* Vol. 62, No. 5, Nov. 1994, p. 67S 2a[1987]2 All ER 826.
3. (1987)2 All ER 826 at 832.
4. As narrated by Peter Singer and Deane Wells in *supra* note 1 at 114–120, quoted in I. Kennedy and A. Grubb *"Medical Law "*(1994) at 821–22.
5. [1985] FLR 445.
6. Id. at 458.
7. (1987)2 FLR 421 (Family Division).
8. (1988) 537 A 2 d 1227 (N.J. supreme Court).
9. 822 P 2d 1314, 4 Cal Rptr 170 (1992).

10. See I. Kennedy and A. Grubb, *supra* note 4 at 822.
11. Sandhya Srinivasan "Surrogacy comes out of the closet" *The Times of India,* Sunday, July 6, 1997.
12. *Supra* note 3.
13. Parker P.J. "Motivation of surrogate mothers: Initial findings" *Am J. Psychiat Law* 1983 140:117.
14. Sister Regis Mary Dunne in her dissent to the Australian National Bioethics Consultative Committee [NBCC] report endorsing gestation contracts quoted by L. Shanner in "The Right to Procreate" *40 McGill Law Journal* 1995, 823-874 at 854.
15. See *supra* note 3, ibid.
16. Judy Stiver's case, see I. Kennedy and A. Grubb, *Supra* note 4, ibid.
17. Ibid.
18. In one reported case, a 26 year old single father in Philadelphia who paid 30,000 dollars to a lady to serve as surrogate mother and bear his child by artificial insemination was charged with beating the five week old child to death. "Death spotlights surrogate parenting". The [Toronto] *Globe* and *Mail* (January 19, 1995).
19. *Supra* note 5 at 543.
20. As quoted by Trevor Allis "Legal Implications of Motherhood by Hire" in *5 Issues in Medical Ethics 1997* (Jan.–March) 21.
21. See *supra* note 9.
22. See Ellen J. Garside "Posthumous Progeny" *41 Loyola Law Rev.* (1996) f.n. 25.
23. Surrogate Parenthood Act, 1988 (Qd) puts a ban on all forms of surrogacy arrangements in Queensland. The Infertility (Medical Procedures) Act, 1984 (Vict) prohibits even non commercial activity by agencies or other intermediaries.
24. See also *A.v.C.* [1985] FLR 445; *Re P Minor Baby M* [Wardship : Surrogacy] (1987) 2 FLR 421 (Fam. Div.); *Baby M.* (1988) 537 A 2d 1227 (NJ Sup Ct).
25. See *Womens Era,* May (First) 1997 at 153.
26. W.J. Winslade, "Surrogate Mothers : Private Right or Public Wrong?" (1981) 7 *J. Med. Ethics* 153.

SECTION II

ELECTIVE BIRTHS

8

AGAINST FEMALE DIGNITY*

Female infanticide — the killing of female infants — was a practice prevalent in India long ago. Though this has now been abolished legally, we find that it is still practised in a different and more "sophisticated" manner and females are being 'killed' even before they are born.

Amniocentesis is a medical technique developed to detect serious mental and physical abnormalities or defects in an unborn baby. If any defect is detected, the foetus can be treated or aborted. Fluid is drawn by needle from the mother's amniotic sac. The cells in the fluid are cultured for diagnosis and tested. The sex of the unborn can also be detected by this test. It may be pointed out here that it is only at a relatively advanced stage of gestation that the foetus is capable of this test viz, around the 16th week of pregnancy.

This test is now becoming quite popular with those who do not want to be burdened with a female child. The question here is how far is it ethical, moral or legal to make use of this test only for the purpose of determining the sex of the child? So far as the use of amniocentesis for detection of mental and physical defects in the unborn child is concerned, it is perfectly in order. Certain defects can be treated prenatally whilst in absolutely negative cases, abortion can be

* *Hindustan Times*, February 15, 1986

resorted to. It is only when the test is made use of for the purpose of determining whether the child in womb is a boy or a girl that problems might arise. What is the right of the unborn child? Is it not discriminatory and against female dignity to have the foetus aborted if it is a female and continue the pregnancy if it is a male? Does if not amount to culpable homicide?

The shorter Oxford English dictionary defines a "person" in two ways, firstly, as "an individual human being" or "a man, woman or child" and secondly as "the living body of a human being". It has been medically established that an unborn child beyond a certain stage of pregnancy has a life and body of it own. By a fiction of law a child *en ventre sa mere* (in the mother's womb) is considered to be an entity for certain legal purpose. Such an entity is given due protection and rights under the criminal law, torts and the property laws as well.

Section 313 of the Indian Penal Code (IPC) says: "Whoever voluntarily causes a woman with a child to miscarry shall, if such miscarriage be not caused in good faith for the purpose of saving the life of the woman be punished with imprisonment of either description for term which may extend to seven years and shall also be liable to fine."

Another relevant section is section 316 which provides, "whoever does any act under such circumstances that if he thereby causes death he would be guilty of culpable homicide and does by such act cause the death of a quick unborn child, shall be punished with imprisonment which may extend to ten years and shall also be liable to fine."

Coming to the abortion law, under the Medical Termination of Pregnancy Act, 1971 (MTPA) where the continuance of a pregnancy involves risk to the life of the pregnant woman or of grave injury to her physical or mental health or where there is a substantial risk that the child that is born would suffer from serious physical or mental abnormalities, then the pregnancy can be terminated under certain conditions. For such termination, the opinion of one medical practitioner, in the case of a pregnancy upto 12 weeks, and of two medical practitioners, in the case of a pregnancy between 12 and 20 weeks, regarding the existence of circumstances (above mentioned) which permit abortion, is necessary.

In the light of these provisions of the I.P.C. and the MTPA, what is the legality of an abortion following an adverse sex determination test? The MTPA definitely does not cover a situation where the gender of the unborn does not happen to be of the choice of the woman. There is no provision for permitting termination of pregnancy simply because it happens to be of a particular sex.

Whatever the effect of carrying a female child on the mental condition of the pregnant woman, surely, the legislators did not provide for a termination on this score alone; and hence such an abortion would be an illegal abortion so far as the Act is concerned.

Similarly, the provision of the Penal Code would be attracted too. Aborting a child upon being dissatisfied after an amniocentesis would not be an abortion or termination for saving the life of the woman and hence would be an offence.

In view of the socio-cultural environment prevailing in our society, a lot of importance is given to a male child. Absence of a male child to continue the family name, business and for the performance of the obsequies, is considered to be a grave curse. Quite often the wife or daughter-in-law is tortured for producing only female children, oblivious or ignorant perhaps, of the fact that it is the male who is responsible for the sex of the child and not the mother.

If resorting to amniocentesis for sex determination becomes common, it is bound to create an imbalance in the male-female ratio of population. This would give rise to problems like rape and illicit relationships, polyandrous marriages would become inevitable, and so on.

Besides being violative of the provisions of the I.P.C. and the MTPA and giving rise to certain other problems, amniocentesis undergone simply to determine the sex of the child and upon detection that the child is a female, having the same aborted, is also against the Constitutional mandate. Article 51A (e) of the Constitution says that it is the fundamental duty of every citizen ". . . to renounce practices derogatory to the dignity of women" and surely this is a practice which is against the dignity of women.

Advertisements and huge hoardings seducing couples/pregnant women to approach them for either determination of the child's sex or for giving such treatment that a male child would be born, are a common sight these days. Though, a family with a mixed male-female population is an ideal gift or blessing but overt acts to "destroy" a particular species does not behove a civilized society. Genetic counselling and medical ethics should discourage this wave. Besides, there is also an element of risk to the mother and the child in the process. Amniocentesis is a medical advancement to detect, diagnose and treat pre-natal defects; it should not become a device to opt for one gender and discard the other.

9

AMNIOCENTESIS AND GENDER DISCRIMINATION IN INDIA*

Introduction

Recent developments in medical science and technology have come both as a boon and a bane. Whilst on the one hand it has brought relief and happiness in the life of people, it's irresponsible handling and misuse, on the other hand, has led to disastrous consequences. One such technology is the use of amniocentesis and other pre-natal diagnostic procedures, like ultra sound, only for sex detection, leading to sex selection.

In many societies, a male child is much more preferred than a female child. According to some reports, China is believed to be the first country in the world to use pre-natal diagnostic techniques for sex detection. Because of the adoption of single child norm, women want to get rid of a female baby, so that they could try again for a male child. This practice is leading to a serious imbalance in the sex ratio.[1] A recent study[a] conducted in 1993 in some provinces in China revealed that the sex ratio was more than 120 boys for every 100 girls. The practice is reported to exist in Korea also. Many

* Paper presented at the International Conference on the Unity of Sciences, Seoul, Korea, August 21–26, 1995

Korean women reportedly, secretly abort female foetuses after finding out their sex. This has created an imbalance in the sex ratio which can also be witnessed in the primary schools. "In the past, a primary school class in Korea comprised 80 pupils — 40 boys and 40 girls and a boy and a girl shared a desk. But during the past few years, a typical class has about 10 to 15 boys with no female partners to share the desk."[3]

According to the same report, many stone Budhas in Korea have been damaged because there is a belief that the Budha's nose has mystical powers to bless women with male offspring. "Women are so obsessed with the desire to have a son that they attack these holy statues and cut off their nose, grind it into fine powder, mix it with water and drink while praying. Noseless statues of Budha can be found in many places."

There is also a mention of "an easier and more non violent manner of prayer for a son at Chikjisa Temple, some 30 Kms. south east of Seoul. The temple, built more than 1,500 years ago is famous for its Pirojon Hall which houses a thousand Budhas made of jade. Each of these statues has a distinctive look and personality, but only one is said to have divine power to bless a woman with a son. She must pin point this special Budha at first glance. This Budha stands out from the other 999. He stands naked, showing off his male 'symbol'."[4]

The Bias against Female Babies

However, in no other country is the medical technology so blatantly misused through a net work of clinics and centres, to discriminate against the female babies, as in India. A son enjoys a preferential status in most of the Indian societies. Having more than one daughter is a curse whereas, any number of sons is welcome. The proverbial blessing to a newly wed bride, "May you be the mother of hundred sons" is indicative of this attitude. A mother with daughters is an object of pity; one having sons is glorified. Even in the family, a daughter-in-law who has sons is pampered and respected, whereas one who has daughters only is treated with indifference. Such pro-male and anti-female attitude can be attributed to the socio-economic and cultural conditions of our society and also to certain religious beliefs, customs and traditions. However, before going into the details of other conditions, it would be pertinent to point out that emphasis on small families is a very strong factor that has given rise to the practice of sex selection. The demographic needs of the country coupled with economic conditions has led to small families with one, two or maximum three children. When families were large, two or three girls in a family of six to eight children did not matter, but now when parents have to restrict the family, they prefer to make a choice. If their target is one or two children, they would not like to take the risk of producing daughters, because daughters are considered to be ornamental whereas sons are a necessity.

A son is a necessity because descent is traced through a male child. In the absence of a male child, the family name becomes extinct. Besides, the last rites and obsequies are also to be performed by the sons only. It is believed that a sonless father cannot attain salvation. Infact the indispensability of a son in the Hindu families led to the custom of adoption — and adoption only of male children. It was only in 1956 that a law was enacted permitting adoption of female children as well.[5]

Besides, male children are needed to help the father in his business and also to take over after the father. Although these days girls are also taking to business but they are not involved in the family business because after marriage they have to move to another family which would mean the business passing out of the family.

Sons are also seen as a source of emotional and financial security in old age. While a daughter after marriage moves away into another family, the son brings home a bride and stays with parents. Though the joint family structure is now fast giving way to nuclear families still the feeling that it is the sons who will look after the parents is very strong. In the absence of old age financial, housing and other security provisions, there is a heavy dependence of parents on sons.

A female child, on the other hand is seen as a burden or a liability. Dowry is one strong inhibiting factor. Parents have to spend huge amounts by way of cash, gifts and jewellery to the girls and also to the bridegroom and his parents. Quite often, if the husband or his parents are not satisfied with the dowry, the girl is harassed. Sometimes she is victimised to such an extent that she commits suicide so that the parents are not burdened with her dowry. Recently there was a shocking report of three teenaged sisters committing suicide by hanging because their father was poor and could not afford dowry for them. Such cases are not unusual.

A girl is also a victim of abuse and exploitation both inside the family and outside it. There is a heavy social responsibility of guarding her chastity. A girl who is sexually exploited is looked down upon. It becomes difficult to get a good marriage partner for her. She is accused of bringing disgrace to the family. Because of the social stigma attached to it, the girl sometimes commits suicide.

Thus, a female child is seen as a source of financial burden and mental tension. Because of all this, people prefer to have boys than girls.

It is significant to note that unwanted female babies were being eliminated even before the advent of pre-natal diagnostic technology. The difference is that earlier they were killed soon after birth whereas now it has become possible to eliminate them even before birth. In fact, it is argued by supporters of female foeticide that this is a kind way of getting rid of female babies. Though in 1870 the practice of female infanticide was banned by law but there are reports[6] of the existence of such practice in certain parts of the country. A study conducted in the South in 1991 revealed that it is a common thing to put female babies to sleep. One woman, aged 30, confessed that, "for the sake of my family, I will continue to kill all female babies that I deliver till I get a boy . . . [after killing her second female baby a few years back] I felt sad, but I had to think of her future which would have been bleak since we could not afford another dowry."

For those who do not wish to have daughters but have compunction in killing a born child, the pre-natal diagnostic technology has come as a relief.

Beginning of the Sex Detection Tests

The beginning of sex detection tests was made in India in the mid-seventies. In fact, it was a fall-out of medical research in reproductive biology, to detect foetal abnormalities with the aid of amniocentesis. When the Indian Council for Medical Research came to know about this aspect of the research it put a ban on these tests. Some medical entrepreneurs, however, were smart and prompt to encash upon the anti-female and pro-male sentiments of our society and allure people. Clinics performing these tests mushroomed and doctors made good business. The first clinic to perform these tests was inaugurated in Bombay in 1977. A circular referred to this test as "a humane and beneficial test". Around the same time another clinic in the city of Amritsar (Punjab) made a big start by putting hoardings around the city. People were attracted with the advice, "Invest Rs..500 now, save Rs. 50,000 later." The message was clear. Rupees 500 referred to the test fee and rupees 50,000 referred to the marriage expenses of a daughter. And not this alone. Some clinics did not hesitate to adopt unscrupulous tactics to attract people. In one clinic, twin female foetuses were kept preserved in a jar for pregnant women to see and imagine that they too could be carrying twin females. The doctors also brought the woman on whom the test was performed so that she could tell other pregnant women how happy she was to be relieved to female twins. The clinic was, reportedly, thronged with long queues of pregnant women, often accompanied by their mothers-in-law.[7] The business ran smooth until in 1982 a wrong diagnosis led to the abortion of a male foetus. The incident created an uproar which snowballed into a national debate. The result of such debate was that a technology which was hitherto known only to some people in certain areas, now became public. People in large numbers started availing the facility of the boy-girl test. The State of Maharashtra had a notoriety of having the highest number of female foeticides.

A group of activists organised themselves to fight this reprehensible practice and in 1986 launched a determined campaign through a forum known as the Forum Against Sex Determination and Sex Pre-Selection (FASDSP). According to a survey conducted in Bombay, 85% of the gynaecologists performed amniocentesis only for sex detection. A private clinic in a Bombay suburb is reported to have performed 15,914 abortions in 1984-85 and 99% of them were preceded by the test. According to an appalling report prepared by a women's centre in Bombay, of the 8,000 abortions preceded by amniocentesis conducted in 6 hospitals in 1986, 7,999 were of female foetuses. There were similar distressing reports from other parts of the country as well.

The public outrage following such exposures pressurised the State Government of Maharashtra to enact a legislation to curb the practice. There was a lot of opposition not only from the general public but also from professionals, bureaucrats and even politicians. The State Government, however, went ahead and passed the law, known as the Maharashtra Regulation of Pre-natal Diagnostic Techniques Act, 1988. The enactment of a legislation with limited application was, however, of no consequence. It only led to an influx of such cases to bordering states which had no law of their own and to which the Maharashtra state law could not extend. Also, for the convenience of those who wanted these tests, courier services and mobile vehicles fitted with necessary equipments, were introduced. In fact, after the Maharashtra legislation, the number of foeticides in the neighbouring Gujarat state increased enormously. The numbers continued soaring in other states too. There were protests all over the country to put a legal ban on the misuse of medical technology to selectively eliminate female foetuses. November 24, 1988 was observed as a national protest day against the practice. Other activities were also organised like signature campaigns, public meetings, rallies and skits, to create public awareness. As a result of this, the central government set up an expert committee to look into the matter and prepare a report. After detailed interviews, surveys and deliberations, the Committee submitted its report in 1989. Based on these recommendations, a Bill was introduced in 1991 on which public opinion was invited. As was to be expected, there was a mixed response with the balance tilting against such law. Notwithstanding all criticism and objectives, the legislators went ahead and passed the Pre-natal Diagnostic Techniques (Regulation and Prevention of

Misuse) Act 1994. Before going into the provisions of the Act, it would be pertinent to discuss the arguments of those who oppose and those who support the law banning the practice of sex selection.

Arguments against the Law

Against Women's Interest

(a) Those who oppose such law argue that putting a ban would be detrimental to the interests of women. If unwanted girls go on increasing, they will be ill treated and exploited. On the other hand, if their number is small they will be more in demand. Their value will go up and there will be lesser crimes against them.

(b) Disallowing a woman a choice in the matter would mean multiple pregnancies. She will go on producing children until she gets a child of her choice. This will have a very bad effect on her health. It will also mean enormous economic burden on the family. Other children in the family would also suffer neglect in terms of maternal care and affection, and access to other needs like food, clothing, education, living space, etc. Thus there would be an overall deterioration in the health and quality of life of the family.

(c) A legal ban would also give impetus to underhand practice. As a result the cost for the test would go up because there will be risk involved for doctors performing it. Whilst those who can afford will get the test and abortion done at safe and hygienically equipped places, others who cannot afford will approach quacks and ill equipped clinics. Thus it will be a great risk to the health and life of women..

Infringement of Freedom

Prohibiting a woman to make a choice in the matter of composition of her family would be an infringement of her freedom. These matters are too personal to be invaded. Besides, if the law gives her the right to space or chose the number of children she wants to have by allowing her to terminate her pregnancy in case of failure of contraceptive device,[8] there is no reason why she should not be allowed to decide on whether or not she wants to have a baby of a particular gender. Moreover, if abortion is legal, it should include abortion of both male and female foetuses.

Against National Interest

An argument that is often advanced in favour of sex selection tests is that it is an effective family planning measure. Putting a ban on sex detection tests for those who want a child of a particular gender would mean unnecessary additions in the family. Thus a couple who wants a son would go on producing daughters till they have a son. In fact it is not unusual to see families with three or four daughters with a son as the last child. If the woman had the option to make a choice, the family would not have had the unwanted girls. Thus, a legal ban on sex detection tests only adds to the problems of over population, with all its deleterious effects on the children, the family and the country, as a whole.

Inefficacy of the Law

Another argument against a legal ban on sex detection and selective abortions is that it would be of no effect. The practice arises out of age old social, economic and cultural conditions of our society which cannot be changed through any law. A woman or a couple who do not wish to have a particular child will in any case get the test done followed by an abortion if the results of the test are unsatisfactory. It requires much less skill to detect the gender of the foetus than any other abnormality. A simple ultra sound is enough. These machines are so common that it is impossible to monitor them. In any case, these need not be registered so, how is any enforcement agency to reach them? Also though medical ethics, and now, even law, enjoins upon doctors etc., not to disclose the gender of the foetus, there is no way it can be enforced. The information can be conveyed by a mere gesture, a nod or a code. Moreover, since abortion is legal, it can be had done anywhere. The nexus between the test and the abortion is impossible to prove. Hence it would only be making a mockery of a law which is unenforceable.

Arguments in Support of the Law

Those who support a legal ban have the following arguments:

Imbalanced Sex Ratio

If the number of females falls, it will create an imbalance in the male-female sex ratio. The census reports[9] already reveal a decline in the number of females as compared to males. It is feared that if this trend continues it would be difficult to find brides for all eligible bachelors, thus encouraging polyandrous marriages. This in turn would increase the reproductive burden on women. Also, there would be more cases of rape, incest and kidnapping.

Maternal Health Hazard

Allowing these tests followed by abortion in case of 'unfavourable' report could mean a number of abortions for the woman. This repeated test-abortion-test cycle until a child of choice is detected is bound to effect the health of the mothers. Indian women are undernourished and anemic, and multiple abortions would only add to their health problems.[10]

Women's Choice — An Anachronism

It is strongly, and rightly, argued that the argument of freedom of choice to a woman in the matter is an utter farce in our context. A married woman is dependent on her husband and in-laws who take decisions. She enjoys no freedom even in the most intimate matters of her life. There are subtle pressures and forces both inside the family and outside it, under which she lives. Hence, there would be no infringement of a woman's right to chose if a law banning sex detection tests is enacted.

Legal Ban: Deterrent Effect

The argument that a legal ban would be completely ineffective is not sound. Agreed, despite legal provisions, certain practices do continue and dowry is the most glaring example, but that is no justification to tolerate something which is improper. Many clinics and doctors who are, or who have been openly, doing the tests would think a couple of times if there is a specific law against it. The fear of bringing bad name or penal consequences will prevent many professionals from performing these tests. And once doctors cooperate and refuse to do the tests, the law is sure to succeed.

General Legal Position

Whatever the socio-economic or cultural conditions of our society be, the legal position is clearly against the practice of sex detection tests simply to get rid of a child of unwanted gender. The relevant laws in this context are the following:

The Constitution of India, 1950

India has a written constitution which was made in 1950. The Constitution is the root of all laws. There are certain fundamental principles laid down in it and all practices, behaviours, rules and laws have to be in conformity with those principles. The use of pre-natal diagnostic techniques for slaughtering female foetuses is a violation of several constitutional provisions. Article 14 guarantees equality before the law, to every person. Article 15 prohibits discrimination on, inter-alia, ground of gender and article 51(A) (e) enjoins upon every citizen, a duty to "renounce practices derogatory to the dignity of women." Selective elimination of the female foetus with the aid of a medical technology is a violation of all the above provisions — it violates the equality clause, it is discriminatory and it is an affront to the dignity of the female gender.

The Indian Penal Code, 1860 (IPC)

The IPC has a chapter on offences against human life. Five sections[11] of this chapter deal with miscarriage. Causing miscarriage is a punishable offence unless it is done to save the life of the mother. Punishments vary depending on the gravity of the offence, e.g. the punishment is higher when miscarriage is performed at an advanced stage of pregnancy, or when performed without the woman's consent or when death is caused by an act intended to cause miscarriage. The only condition which permits miscarriage without attracting penal consequences is when it is done to save the life of the mother. However, this law could not prevent pregnant women from going to quacks for an abortion. Operations were performed by unqualified midwives or quacks under unhygienic conditions thereby risking the health and life of women. In order to remove these difficulties, the Parliament enacted a special law in 1971, known as the Medical Termination of Pregnancy Act.

The Medical Termination of Pregnancy Act, 1971

The Act permits a medical termination of a pregnancy on the following grounds.

(i) where continuance of pregnancy involves risk to the life or to the physical and mental health of the mother,[12] or

(ii) where there is substantial risk that if the child is allowed to be born, it would suffer from such physical or mental abnormalities as to be seriously handicapped.

The operation for termination of pregnancy can be performed only by a registered medical practitioner, and where the length of the pregnancy does not exceed twelve weeks. If it is over twelve weeks but within twenty weeks, then it has to be performed by two registered medical practitioners. The doctors who perform the operation must be satisfied that there are justifiable grounds for the same.

Simply because a foetus is female is no ground to terminate a pregnancy. Femaleness, *per se,* is not a disease to justify an abortion unless a sex-linked disorder is indicated. Besides, no pregnancy beyond twenty weeks can be legally terminated unless it is essential to do so to save the life of the woman. And amniocentesis, which is one of the methods used for sex detection is normally carried out in the sixteenth week of pregnancy. The time gap between the test and the final report could be another 3 to 4 weeks or may be even more if the test has to be repeated. Thus, by the time she decides to have an abortion, it is beyond the statutory period of 20 weeks.

However, there are several loopholes in the Act which facilitates the misuse of the law purely for selective elimination of the female foetus. The most glaring is the provision[13] which allows abortion when pregnancy occurs as a result of failure of contraceptive device used for the purpose of limiting the number of children. This provision is used as a scapegoat. The woman gets the test done at one place, then goes to another place for abortion, and the plea invariably taken is the failure of contraceptive device. There is not way to verify this statement.

There is yet another clause which permits abortion on the ground that continuance of the pregnancy would involve a "grave injury to her physical and mental health" which has a great potential for misuse. This is further elucidated by adding the following explanation:[14]

In determinating whether the continuance of a pregnancy would involve such risk of injury to the health, account may be taken of the pregnant woman's actual or reasonably foreseeable environment.

In our society the birth of a female baby is an occasion to console rather than congratulate, especially if she is not the first. The life of a woman who gives birth only to daughters is miserable because she faces harassment in the family and embarassment in the society. Sometimes, excuses are found to divorce her so that the man could bring another wife who might give the family a male child. In this situation, can she take a plea that continuing with pregnancy without confirming the sex of the foetus would cause injury to her mental health? The fact is that many doctors who perform abortions after amniocentesis or ultra sound tests feel fully justified. They argue that in the conditions in which many Indian women live, it is better for them that a female foetus be aborted. In fact sub-section 3 of section 3 of the Act provides a great protection to the supporters of sex detection tests and gives credence to their argument that an unwanted child should not be thrusted on the parents.

Also, the Act gives wide discretion to the doctor by exempting him from any liability if he performs the abortion in "good faith". The term good faith is very subjective and even abortions which fall foul of the provisions of the Act cannot be punished. Section 8 gives a garb of protection to the doctors. It is an ouster clause which says that no suit or other legal proceeding shall lie against any doctor for any damage caused or likely to be caused by anything which is done or intended to be done in good faith under the Act.

The Rules under the Act provide that the doctors performing the abortion have to certify their opinion in a prescribed form as regards the necessity of such pregnancy termination. This, however, is just an eye wash. The grounds are specified in the form and numbered (i). . . (ii) . . . (iii) . . . (iv). . . (v) and the doctor has just to tick mark against it and state that he was of the opinion formed in good faith that termination of pregnancy was necessary. No further details are required, like, what type of injury to the mental or physical health of the mother would have been caused if the pregnancy were not terminated, or what type of risk of abnormality in the child was apprehended, or what was the device or method of family planning which was adopted and which failed, resulting in the unwanted pregnancy, and so on. The whole process is, in fact, as good or as bad as abortion on demand.

The Pre-natal Diagnostic Techniques: (Regulation of Misuse) Act, 1994

The various escape routes in the existing legislation coupled with weak enforcement conspired with the prevailing conditions and facilitated mushrooming of sex detection centres and female foeticides. A stringent all India legislation in the matter was seen as the only viable solution to curb the practice. Hence, our Parliament passed the special law.[15] The following are some of its salient features:

This is an all India legislation unlike the Maharashtra or other State Acts which are applicable to that particular state alone. Under this Act, the pre-natal diagnostic techniques may be used only to detect "genetic or metabolic disorders or chromosomal abnormalities or certain genetic malformations or sex-linked disorders."

There is a provision for compulsory registration of every genetic clinic, laboratory or centres with minimum standards prescribed in terms of space, equipment, and qualifications of the staff.

Not all pregnant women are eligible for the test. Only those who fall in the risk category as specified by the Act, can have a pre-natal diagnostic test performed. The conditions are — the woman must be above 35 years; history of two or more abortions or foetal loss; exposure to potentially teratogenic agents, such as drugs, radiation, infection or chemicals; history of mental retardation or physical deformities such as spasticity or any other genetic disease; or any other conditions as may be specified by the Supervisory Board.

The whole purpose behind this legislation being to ban sex detection tests, there is a specific provision to that effect. Even advertisements in any form, of facilities for the tests, are prohibited .

In order to look into the policy and implementational matters, there is a provision for setting up of various bodies. Their composition, powers and functions are also clearly defined in the Act.

Penalties have been provided for misuse of diagnostic procedures. Any medical geneticist, gynaecologist, medical practitioner or any person who owns a clinic or renders his professional or technical service at such clinic shall be liable, on conviction, to imprisonment upto three years and fine upto Rs. 10,000, if he contravenes any of the provisions of the Act. Likewise any person who seeks the aid of a centre or laboratory or clinic for carrying on a pre-natal sex detection on a pregnant woman for any purpose other than those specified under the Act, shall also be liable to similar punishment.

In the case of a registered medical practitioner, the Medical Council may even suspend his license for two years for the first offence and permanently, if the offence is repeated.

In the case of a woman, unless the contrary is proved, there is a presumption that she has been compelled by her husband or the relatives, to undergo the test and so he/they will be held liable for abetment of the offence.

It would be relevant to note that the issue of the woman's liability in such cases was debated upon at length. An argument was given that the Indian woman is ignorant as well as helpless. She has no choice in the matter except to surrender to the desires of her husband or mother-in-law, or there might be other subtle but grave pressures.[16] In such situation, it would be very unfair to hold the woman responsible. The other side of the argument was that if a woman were to be totally exempted from any liability, she would be made a scape-goat by the husband or other relatives who in fact would be the culprits. Thus no one would be penalised for the illegal act in such cases. Hence, the legislators found the via media.

Conclusions and Suggestions

The law against sex detection tests and sex selection will become operational sooner or later but the real question is, will this alone be effective enough in banning the tests? The answer quite clearly is in the negative. Despite a ban by the Medical Council of India, these tests are being performed with impunity. One can see advertisements about facilities for these tests at many places. There is a recent report of how a reporter of a popular daily caught a doctor performing these tests.[17] Posing as a pregnant woman she approached him for the test. She told him that her mother-in-law wished to have a male child and so she would like to go in for the test. The doctor took his fee for the amniocentesis test, gave a receipt and asked her to come for the test on the following day. When the lady enquired whether there were any side effects of the test, he reassured her that it was absolutely safe and stated further that he had already conducted a few of these tests during the day.

According to a very recent survey conducted by the Department of Anatomy at a medical college at Jaipur (in the state of Rajasthan) 3,500 female foetuses are killed every year after the test in that city alone. The report states that the use of widely available ultra sound facilities for the determination of the gender of the foetus eliminates ten female foetuses daily at three medical centres in the city where these tests are carried out.[18]

There are also reports of professionals conducting crash courses to train doctors in pre-natal testing. Such training is obviously to be used for conducting sex detection tests since resort to pre-natal diagnostic procedures for anything other than this is almost unknown.

The practice thrives because even the doctors find nothing wrong in it. As pointed out by one doctor[19]:

Today, a woman with more than one daughter has a gun pointed out at herself and her progeny. How can you deny her the right to have a son instead of a third or fourth daughter? You can't wish away centuries of thinking by saying that boys and girls are equals.

According to another well known Bombay based gynaecologist,[20]

> Until the whole Asian community changes where a male issue in the family is a must, we as scientists can help out the poor mother who year after year produces a baby until a boy is born . . . It is a very peculiar situation. If you don't do it you are creating an unhappy situation for the mother and the child. If you do it, you are discriminating on the basis of sex.

The general feeling amongst the doctors and others who support the practice is that it is much more cruel to bring her into a world where she is unwanted, and make her suffer throughout life than eliminate her at a stage where there is hardly any format.

Whatever the alibi for tolerating the practice, it cannot be denied that it is deplorable. It has gained notoriety to such an extent so as to make people wonder whether at all India has a right "to take the newest technology from the west and use it for something as reprehensible as the slaughter of female babies."[21]

As pointed out above, law alone cannot solve the problem. What is needed is a change in the whole structure and thinking of the society where girls and boys will be viewed as equals. Customs like dowry make a girl a financial burden on the parents, so it is dowry which has to be fought and eradicated. Girls who do not satisfy the demands of greedy husbands and in-laws are tortured by them. What is therefore needed is a very stringent social and legal action to punish such people. There is a tremendous need to create awareness against these evil customs and practices. Education is one sure way to bring change in social attitudes. If girls are educated and aware, they will fight back and not suffer abuse, exploitation and torture, which is one of the reasons why parents hesitate to have girls.

Financial independence is another important measure to bring about change in the status of females. They can thus be as much an asset to the family and society as boys. It is common knowledge that girls who are illiterate and financially dependent on their husbands are more oppressed. They feel that they have nowhere else to go and so keep suffering. With financial independence, they can lead their own life and live in dignity.

Law can go a long way in conferring equal rights and status on girls. Infact, we have many special and general laws for the welfare of women. The succession laws give equal inheritance rights to sons and daughters. The matrimonial laws give equal rights to the husband and wife and under the employment laws, equal job opportunities are given to males and females.

Our government has also adopted many measures and schemes to fight out discrimination against girls. The television and radio are being extensively used for conveying a message that girls and boys are equal. Almost daily there are advertisements and telefilms to educate the people. Girls are encouraged and given equal opportunities for education and employment. Some states have also introduced schemes for the welfare of female children — like state adopting unwanted female babies and financial assistance to parents to meet the education, medical and marriage expenses of girls. There are also schemes to provide free education to girls with additional incentives. However we still have a long way to go.

In discouraging sex selection the medical profession has an important role to play. It must inform the public about the risks involved in the test. If informed about just a single risk of losing a male foetus in the process of the test, many women would refrain from the test. Also, if doctors refuse to perform the test, the practice will automatically cease. However, simply banning a test and thrusting female children only to be neglected, abandoned and ill treated would be a remedy worse than the evil. Conditions conducive to the dignity of females have to be created. Awareness amongst the people that girls are as much an asset as boys, equal rights under the law and a stringent legislation to ban the practice, all combined, alone can help in bringing a change.

REFERENCES

1. Shanghai Evening News reported in *Indian Express* (8.12.1988).
2. "Disappearing Girls", *Asiasweek* (3.3.1995) as quoted in *Competition Success Review* (April, 1995) at 115.
3. Roe Bong Mi, "The 'Sun' of the family", *Hindustan Times*, Sunday Magazine (22.2.1987).
4. Ibid
5. The Hindu Adoptions and Maintenance Act, 1956. (As its very name conveys, the Act is applicable only to Hindus).
6. "Born to Die", *India Today* (June, 1986) which estimated that 6,000 female babies had been poisoned to death during the preceding decade in the district surrounding the town of Madurai in Tamil Nadu. See Elisabeth Bumiller, *May You Be the Mother of a Hundred Sons* (1991) at 104–112; See also Usha Rai, "Female Infanticide Thrives in South", *Indian Express* (8.8.1992); Pushpa Iyenger, "Girls in Salem are born to die", *The Times of India* (30.8.1992).
7. *Facets*, vol. 1, No. 3 (August, 1982).
8. Explanation II of Section 3(2)(ii) of the MTPA.
9. Appendix I.
10. In India, maternal mortality rate is said to be the second highest in the world with 400 to 500 per 100,00 live births — and 22% of these is reportedly due to abortion.
11. Sections 312 to 316 of the IPC.
12. Pregnancies which are the result of rape or failure of contraceptive device are treated as situations causing injury to mental health of the woman, vide Explanation I & II of Section 3(2). See Appendix II.
13. Explanation II of Section 3(2).
14. Section 3(3).
15. The Act has been passed and gazetted but the rules are still under formulation. Hence the central law has yet to take off. There are some states which have their own laws banning such tests but they are applicable to those states only. However, once the Central Act becomes operative, the state legislations will stand repealed.
16. For instance, her marriage might be threatened if she gives birth to a female child; or the child if born, might be neglected or abandoned.
17. Swati Chaturvedi, "Boy or Girl? They will tell you for a price", *Statesman* (22.5.1994).
18. "Sex-tests kill 3,500 foetuses in Jaipur", *Indian Express* (31.1.1994).
19. Dr. Datta Pai, former Director, Family Planning Hospital in Bombay.
20. Dr. Soonawala.
21. Elisabeth Bumiller, *May you be the Mother of a Hundred Sons* (1991) at 118.

APPENDIX I

All India Sex Ratio 1901–1991

Census Year	Sex ratio per 1,000 males
1901	972
1911	964
1921	955
1931	950
1941	945
1951	946
1961	941
1971	930
1981	934
1991	929

APPENDIX II

Extracts from the Medical Termination of Pregnancy Act, 1971

3. When pregnancies may be terminated by registered medical practitioners: (1) Notwithstanding anything contained in the Indian Penal Code, a registered medical practitioner shall not be guilty of any offence under that Code or under any other law for the time being in force, if any pregnancy is terminated by him in accordance with the provisions of this Act.

(2) Subject to the provisions of sub-section (4), a pregnancy may be terminated by a registered medical practitioner:

(a) Where the length of the pregnancy does not exceed twelve weeks, if such medical practitioner is, or

(b) Where the length of the pregnancy exceeds twelve weeks but does not exceed twenty weeks, if not less than two registered medical practitioners are, of opinion, formed in good faith, that:

(i) the continuance of the pregnancy would involve a risk to the life of the pregnant woman or of grave injury to her physical or mental health; or

(ii) there is a substantial risk that if the child were born, it would suffer from such physical or mental abnormalities as to be seriously handicapped.

Explanation I. Where any pregnancy is alleged by the pregnant woman to have been caused by rape, the anguish caused by such pregnancy shall be presumed to constitute a grave injury to the mental health of the pregnant woman.

Explanation II. Where any pregnancy occurs as a result of failure of any device or method used by any married woman or her husband for the purpose of limiting the number of children, the anguish caused by such unwanted pregnancy may be presumed to constitute a grave injury to the mental health of the pregnant woman.

(3) In determining whether the continuance of a pregnancy would involve such risk of injury to the health as is mentioned in sub section (2) account may be taken of the

pregnant woman's actual or reasonably foreseeable environment.

(4)(a) No pregnancy of woman, who has not attained the age of eighteen years, or, who, having attained the age of eighteen years, is a lunatic, shall be terminated except with the consent in writing of her guardian.

(b) Save as otherwise provided in clause (a) no pregnancy shall be terminated except with the consent of the pregnant woman.

10

[TIME] ELECTIVE BIRTHS*

Faith in astrology and superstitious beliefs guide the lives of a large section of our people in diverse ways. It has led to various crimes and anti-social acts. And now such beliefs, in connivance with medical science, are prompting man to play God by choosing the time of birth of babies. Thus both death and birth can now be pre-poned or delayed with the help of medical science and technologies. While I plead ignorance of whether people have started to choose the moment of death of their near and dear ones to ensure their entry to Heaven, time-elective births are very much in the news.

There are reports of more and more expectant mothers being guided by the stars to decide their delivery time. Astrologers are making money as they are being consulted to determine the'holy' and auspicious time for the entry of the child into the world. And the astrologers believe that the time of delivery is indeed significant in influencing a child's destiny. "If instead of being born in the morning a child is delivered in the afternoon, the planetary positions change. Thus fate also changes accordingly," observed one astrologer.

It is not only the time of delivery but the time of conception as well (aadhan) which is considered to be significant.

---

* *Hindustan Times*, June 11, 1997

We find support for this in the Indian mythology too. The birth of Ved Vyas, the great scholar and author of Mahabharata is believed to have been the culmination of a union at a time when planetary positions were conducive to produce a philosopher.

Gynaecologists and obstetricians are being requested to perform the deliveries on particular days or hour. Reportedly, a lady whose delivery was calculated in the third week of April got delivered on Baisakhi day, i.e., April 13, as it is considered to be an auspicious day. Likewise, a devotee of Sai Baba got her delivery time to coincide with the birthday of her Guruji. Cases of such time-elective births are on the increase. It is argued that if people can get married by consulting horoscopes, what is wrong in having babies at an auspicious hour?

One, however, fails to understand the analogy. Child birth is a natural, biological, medical phenomenon which has its own course — unlike marriage ceremony. Interfering with the course of nature can be hazardous for the mother as well as the child. It would also mean caesarian deliveries to keep to the desired 'auspicious' day and time when natural normal child births are possible. While some gynaecologists are against the practice, it being medically hazardous and unethical, others feel that there is no risk in advancing (or postponing) pregnancies by some days if the condition of the mother and baby permits so. The fact of the matter, however, is that caesarian deliveries are on the increase and though this may be in the interest of the doctors and nursing homes — commercially and as a matter of convenience — it cannot be denied that natural normal delivery is safer. It is unethical as it interferes with the course of nature. Man is trying to play God by deciding the time of delivery.

SECTION III

FOETAL RESEARCH/PROTECTION

11

TRADING IN FOETUSES*

Education, technology and scientific developments have changed the complexion of human life, living and even thinking. What was supposed to be unthinkable, and impossible once is a reality today. It cannot however be denied that human ingenuity is a double edged weapon which can be both defensive and offensive. In other words, it has contributed towards attainment of miracles and lofty goals on the one hand but at the same time given rise to certain ignominious activities and deals as well. We are living in an age and environment where everything is commercialised — animate and inanimate, alive and dead, still-born and born.

An activity that is still abstruse, but is very much in the air, could shock the conscience of many. This is the use and abuse of , or trading in, the human foetus. Live animals have been an object of experimentation and medicines, dead human bodies serve medical science by being an object of research and experimentation, and also provide certain organs which can be transplanted to the living human beings; but that embryos could be commercialised for medical or cosmetic purposes was unthinkable.

Now that abortion has been legalised and liberalised in many countries, including ours there is no dearth of foetuses or embryos. It is believed that there are a few embryological

* *Hindustan Times*, January 1, 1984

laboratories abroad where researches on foetuses of different ages are conducted. After an abortion is done the foetus is preserved, dissected or artificially sustained for experimenting or for use in preparing medicines or cosmetics.

It is submitted that the use of foetuses for such purpose is bound to give rise to various issues like legal, religious, moral or aesthetic. It may be argued that if cadavers can be used for research and experimentation, why not the foetus which, though technically alive at times, has no chances of survival whatsoever. The former is a human body which has lived its life and now is devoid of life and the latter is a condition where the withering process takes place before life comes into actual existence.

It cannot at the same time be denied that such practices need to be strictly regulated. It is not improbable that contracts might be entered into with women to conceive and then undergo an abortion at the right period (when value of foetus for use is the maximum) at a high price. When contracts for hiring wombs are very much around where a woman carries a child for someone else to hand it over to him or her after delivery, there is every likelihood that as and when the practice of using foetuses becomes popular, many commercial firms, doctors or hospitals might be on a look out for women who could do the job for money.

Agreed, such contracts would not stand the scrutiny of law being immoral or anti-social etc., but it is common knowledge that many things and activities do thrive despite being neither legal nor moral. The legislators and administrators should foresee the issue, its ramification and implications before it becomes unsavory. Appropriate provision should be made in the Medical Termination of Pregnancy Act, 1971, and in the Indian Penal Code to ensure than no abortion is resorted to with an ulterior motive. Disposal of the foetus should be strictly regulated and its commercialisation should be made a serious offence.

12

OF FOETAL RIGHTS*

A pregnant woman takes tetracycline during her pregnancy resulting in the discoloration of the teeth of the child born subsequently. The child sues the mother successfully in a Michigan Court (*Grodin* v. *Grodin*, 1980).

A child is born defective; a pregnant woman who had prior knowledge of the probability of its being born so may be sued by the child. (*Curlender* v. *Bio-Science Laboratories*, California, 1980).**

Not fictions but real cases indeed! One is led to wonder whether there is any limit to the extent of claims and liabilities and that too by an entity which is human only by a fiction of law. It was only a few decades back that an unborn child was 'given' certain legal rights against third parties and now we have claims of foetuses against their mothers for pre-natal injuries.

In one case an American court has allowed a child to sue a pharmaceutical company for personal injuries which were the result of damage to its mother's chromosomes even prior to conception (*Jorgensen* v. *Meade-Johnson Labs*, 1973).

---

* *Hindustan Times*, June 27, 1986

** The court, in this case (1980, 165 cal Rptr 477) saw "no sound public policy which should protect those parents from being answerable for the pain, suffering, and misery which they have wrought upon their offspring". (Dictum.)

An unborn child is also vested with certain property rights. A child *en ventre sa mere* i.e. in the mother's womb, is considered to be a 'child' and properties can be bequeathed to it.

The criminal law also recognises the existence of an unborn child in various ways. Doing away with the foetus or a quick unborn child is considered to be a crime. Also under the abortion laws, termination of a pregnancy beyond a certain period is considered to be an offence.

What is, however, most thought provoking is a development in certain parts of the world which concedes legal rights to a foetus *qua* foetus against the mother. A foetus is completely dependent on and inter-related with the body of the woman and virtually every act of the pregnant woman has some affect or the other on it. There would be no end to her civil and criminal liabilities — should such liabilities exist — for pre-natal injuries. Let us examine some of the very mundane things which might have adverse effects on the foetus.

A woman who does not eat properly harms the foetus. Sub-standard nutrition during pregnancy is said to result in low birth weight which causes many defects. It is believed that, "nutrition is the single most important exogenous influence in the life of the foetus".

Women who smoke or drink during pregnancy might be doing a lot of harm to the unborn baby. "A single cigarette smoked by a pregnant woman, can disrupt the foetus' heartbeat", according to a medical report. Consumption of alcohol may result in "foetal alcohol syndrome" which consists of "growth retardation and assorted congenital defects affecting other organs."

Coming to diseases and medicines — diabetes in the mother may cause cerebral palsy with mental retardation; congenital syphilis may cause blindness, retardation and birth defects. Likewise common diseases like malaria, chickenpox, measles and rubella have potential adverse effects on the development of the foetus.

Simple medicines taken during pregnancy could also prove harmful. Cough medicines may cause congenital goitre or skeletal liver or brain damage; antacids and laxatives may cause kidney and brain damage; quinine and its derivatives may cause deafness. Even such a simple and common drug as aspirin may cause damage to the nervous system, kidneys and liver of the child-to-be.

Apart from all this, exercises and sexual life of a pregnant woman has a very great effect on the foetus. According to Dr. Richard Naeye of the Department of Pathology of Pennsylvania State University, "a pregnant woman's engaging in sexual intercourse is more dangerous to the foetus than the combined effects of her use of alcohol and cigarettes . . . due to a bacterial infection known as chorioamnionitis which is apparently transmitted to the womb by semen".

A working woman is exposed to various hazards—especially those working in chemical industry and drycleaning establishments. It is believed that the fertility or child bearing capacity of certain women is impaired because of their working in certain postures or exposure to certain influences. Women who work at home are also exposed to certain hazards like household accidents, falls or burns etc., which might effect the foetus.

In one case (*Commonwealth* v. *Cass, Massachusettes* 1984) it was held that a viable foetus is a person for purpose of state's vehicular statutes and therefore a woman whose reckless driving results in the loss of her late pregnancy may be subject to prosecution for homicide. In some American states (like Connecticut and North Carolina) a woman could be held liable to her foetus even if she rides with a driver whom she knows to be intoxicated. Her liability would possibly also extend for injuries caused to the foetus in the course of alighting from and getting into an automobile.

In some American states a woman can be deprived of custody of her child before its birth if it is established that her actions during pregnancy endanger the foetus. To mention a very recent case, a court seized custody of a foetus—in other words, the pregnant woman—and made it a ward of the state because of "abuse" by its mother who had the habit of taking heroin (Chicago, 1984).

There have also been cases where courts have ordered women to submit to blood transfusions in the interest of the foetus. Cases compelling women, against their wishes, to undergo caesarian delivery, instead of normal, are also known.

Any rights granted to the foetus against the woman should ensure that the autonomy of the mother is not unduly impaired with intrusions on her body, life and activities. A woman who choses to carry her pregnancy through would, in all probability, be deeply concerned about the welfare of the child. Creating a "liability" on the mother and a "right" on the foetus is also bound to create problems of proof of identifying causal connection of her acts which have allegedly cause injury.

What is needed is the awareness on the part of a woman of potential risks or dangers of her habits and behaviour on the foetus and the necessary steps to be taken to combat the same and prevent pre-natal injuries. Women also need to be educated on their "duties" during this phase of their life like abstention from certain foods and activities, regular pre-natal check-ups etc. The interest of a foetus and the woman should not be viewed as hostile, with the woman being subjected to a tension of being constantly policed by the state of what she ate and drank, what physical activities she engaged in, the quality or frequency of her marital life, and so on.

13

A MOTHER'S PREROGATIVE?*

Medical advancements and refined surgical techniques have improved foetal survival rates but at the same time it is likely to give rise to conflicts and various legal issues. Since surgery on the foetus can be conducted only through surgery on the mother, difficulties may arise if the pregnant woman refuses to consent to the intervention.

In fact, the last few years have witnessed a rise in the incidence of judicial intervention by way of imposing restrictions or ordering medical procedures, including blood transfusion, caesarean surgery, etc., on a pregnant woman, in order to save the life and health of the foetus she is carrying.

The use of drugs, alcohol, tranquilizers, heroin, tobacco, etc. have an adverse effect on the foetus. For example, chronic alcoholism may lead to foetal mal-development, commonly referred to as foetal alcohol syndrome. Smoking results into lower birth weight and impairs growth. Courts have intervened to impose restrictions on the use of such things by the mother. Intrusions by way of operations etc. have also been ordered to save the life of an unborn child. For example, in *Grodin* v. *Grodin* (Michigan, 1980), a mother was held liable in court because she used tetracycline during her pregnancy which led to discoloration of the child's teeth. This was viewed

* *Hindustan Times*, March 11, 1990

by the court as an interference with a child's "legal right to begin life with a sound mind and body".

According to one report, a child was born to a woman who used drugs during pregnancy despite medical advice not to do so. The child was born with severe disabilities and died within two months. Criminal charges were framed against the mother for failing to provide proper medical care for the child and ignoring the doctor's advice (Pamela Rae Steward's case, San Diego, 1986). Likewise, in a recent decision by the District of Columbia Superior Court, a pregnant drug addict, who was convicted of theft, was ordered to be incarcerated until the conclusion of her pregnancy because of her inability to control her drug addiction. (*U.S.* v. *Vaughn*, 1988).

Intrusions into the woman's body and life are severer when direct medical procedures are ordered on her in order to save the foetus. In *Raleigh Fitkin, Paul Morgan Memorial Hospital* v. *Anderson* (U.S. 1964) the New Jersey Supreme Court granted a hospital's request to force blood transfusions on a pregnant woman who refused the treatment for religious reasons. These transfusions were considered necessary since there was a risk that the woman would haemorrhage at some point and both she and the viable foetus would be at risk without blood transfusion.

A lady in the 39th week of pregnancy was diagnosed as having complete placenta previa, a condition in which the doctors estimated the survival chances for the foetus at 1 per cent and for the mother at 50 per cent if caesarean is not performed. The lady refused to undergo caesarean. The hospital sought a court order under which surgery could be performed even against the mother's wishes. It was held that "the State's interest in the potential life of the unborn living human being overrides the mother's religious beliefs (*Jefferson* v *Griffin Spalding County Hospital Authority*, 1981).

Judicial intervention was also sought against a pregnant woman to protect foetal rights in *Taft* v *Taft* (Massachusetts, 1983). The lady was enjoined to undergo a surgical procedure to close her cervix to prevent a probable miscarriage. She wanted the child but objected to the operation on religious grounds. While the High Court stated that there was no State interest which justified the curtailing of the woman's constitutional right of privacy and refuse treatment, the Supreme Court of Massachusetts vacated the ruling and ordered surgery.

In this context, a reference may be made of a very recent case which had a tragic end. In *re A.C.* (Columbia, 1987) a competent but terminally ill woman was 26 weeks pregnant when a tumour was diagnosed in her lung. She was given just a few weeks to live. After discussion with her doctors and family, she agreed to take a treatment that might extend her life until the time the foetus was 28 weeks old (chances of viability being greater at that point). She, however, stated that her care and comfort were to be ensured by the doctors before that of the foetus. Her family members and physicians agreed with this course of action.

When this plan was communicated to the George Washington University Hospital administration, advice was sought from the counsel for the university. The counsel sought a court order for a caesarean. Within six hours the trial court appointed attorneys to represent A.C. and the foetus and a hearing was held at the hospital.

The court found that the foetus was viable and that "there was a State interest in protecting the potential life of the foetus". Initially A.C. agreed to the procedure but a few minutes later she refused and when the surgery was scheduled, a telephone appeal was made for a stay. The request for stay was turned down and surgeons performed the caesarean the same evening. The foetus was not viable and survived for only two hours; A.C. died two days later.

The court, while; recognising that "we all know that we may have shortened A.C.'s life span by a few hours" sought to justify its order by discussing "the importance of balancing between the individual's right to bodily integrity against the State's interest in potential human life".

It was also remarked that it wasn't the case that "A.C.'s good health was being sacrificed to save her child's interest, although her condition was clearly affected", thereby conveying that since A.C. already was in a serious state, there wasn't much at stake so far as her health and welfare were concerned. One wonders if terminal illness could be equated with "non-personhood" and the interest of a living person could be made subservient to the interest of a life-to-be!.

Thus we find that medical technology, with all its miracles and blessings is heading towards serious intrusions and encroachments in the most intimate and personal domains of a woman's life. The courts are faced with a dilemma in choosing between a child's right to be born free of injuries and a woman's right to control her own body. The various issues that arise in such cases are—what is the status of the unborn child; what is the extent of a pregnant mother's duty in respect of the foetus she is bearing; can the law impose restrictions on the habits, life-style and other activities of the woman; would intrusion into her privacy and autonomy not be an infringement of her right to life and personal liberty guaranteed under the Constitution; does the State have a duty to protect the potential life of an unborn and if so can it overrule the objection of the woman to such intrusions?

These and various other sensitive and intricate issues need to be carefully examined and resolved. In any case, there cannot be any absolute rules and each case has to be considered individually. There is no doubt that foetal health is a State interest—for unhealthy and deformed children are a heavy financial burden on the society (apart from the pain, misery and emotional shock they suffer and cause to the near and dear ones). But at the same time there is a need for a proper respect for the autonomy and privacy of a woman. The protection of the foetus with minimal intrusiveness has to be achieved. There has to be a proper balancing but if unfortunately the situation is such that a tilt is needed, then it should be in favour of the existing rather than the one-to-be.

14

A MEDICO-LEGAL DILEMMA*

Medical and technological advancements over the past few decades have brought about a revolution in human life and society. So fast is the pace of such advancements that the law finds itself in a dilemma when a medico-legal issue crops up.

A development in the field of medicine which is bound to have far reaching social and legal implications is research on the possibility of the foetal tissue transplantation. Since the 19th century, research on foetal tissues has been going on. Foetal tissues are obtained at different stages of gestation for the purpose of studying their bio-chemical properties and use in treatment for diseases. In fact in 1954, the Nobel Prize for medicine went to researchers who used cell lines created from human foetal kidney cells to develop the polio vaccine.

Such research holds promises of treatment for various other diseases too like diabetes, Parkinson's disease, leukemia, epilepsy, paralysis etc. The decision in *Roe v. Wade* (1973) and the controversy over abortion and foetal research that followed led the American Congress to set up in 1974 the National Commission for Protection of Human Subjects of Biomedical and Behavioural Research. The Commission has laid down certain regulations which explicitly ban the investigator from any sort of involvement in the matter of pregnancy or

* Hindustan Times, July 29, 1990

abortion. It also says, "activities involving the dead foetus, macerated foetal material, or cells, tissue, or organs excised from a dead foetus shall be conducted only in accordance with any applicable state or local laws regarding such activities".

There are various bio-technology companies which are carrying on research and experiments on foetal cells. The National Institute of Health in America, reportedly spent 11.2 million dollars in 1987 on research on the use of human foetal tissue. In Hana Biologics Inc., in Alameda, California, researchers have developed a process by which they identify and isolate cells from the pancreas of an aborted foetus. These are then purified, proliferated and reconfigured, making them suitable for transplantation.

The company has already tested foetal tissue transplants on 25 diabetic patients. Research and experiments for curing Parkinson's disease are also being conducted. In view of the developed research processes that cause the cells to proliferate in the laboratory, the amount of foetal tissue needed is decreasing.

There are several tissue banks which collect foetal tissues and other human tissues. Foetal tissues are obtained by researchers also from the National Disease Research Interchange (NDRI), a foundation funded by the National Institute of Health, and from the International Institute for the Advancement of Medicine. Both are non-profit foundations.

As it happens in case of every new dramatic research and advancement, many critics of foetal research and transplantation also feel that "by allowing science to proceed, we may unleash an irrepressible force that may brutalise society, leading to commodification of and marketing in human flesh." It may be recalled that two decades back when heart transplantation technology was making a headway, its critics "prophesied the inevitable brutalisation of society".

The anti-abortionists are specially against foetal research. They hold that law should prohibit all benefits gained from abortion because they fear that wide spread appreciation of abortion might increase public acceptance of the same. Another argument advanced by them is that in case of women who are ambivalent about abortion, the knowledge about potential benefit of abortion might tip the balance in its favour. This argument, however, does not hold ground for, as pointed out by one researcher, if that be so, society should also ban the transplant of organs from suicide victims who are frequently ambivalent about the decision to end their lives. The knowledge of a potential benefit viz., the use of his organs for transplant—and in this case the donee can be specially named—might tip the balance in favour of suicide. No such ban, however, has ever been suggested.

Another argument by the opponents is that it will induce women to conceive and abort thus resulting in wholesale "tissue farming by women . . . growth of human husbandry and industrial breeding factories". Some fear that the technology would trigger the start of a "brave new world in which one generation rejuvenates itself with fresh tissue from the next". Recalling Dante's *Inferno*, one critic asks, "Into which circle of hell would these masters of human vivisection via foetal parts transplantation be placed?"

Many associations have joined hands to protest against this research and practice. The International Association to Combat the Exploitation of Human Foetuses held a symposium in Paris a few years ago and pleaded for its ban.

It cannot be denied that some of the fears of the opponents are not entirely unfounded. The risk of commercialisation and exploitation is there. In one report in *Le Monde*, a prestigious newspaper in France, there was an indication that "some Korean women had agreed to supply serviceable material to a US army laboratory in Maryland".

There is also a report that a customs official intercepted a truck load of human foetuses on the Swiss-French border. The cargo was destined for a French cosmetics firm making skin rejuvenating products.

It is not unlikely that some women in financial distress might be exploited or tempted to conceive and abort to provide the "material" for research and medical use. These risks and fears, however, can be mitigated and combated by regulations, safeguards and statutory provisions. The Medical Termination of Pregnancy Act and the Penal Code can have provisions ensuring that no abortion is resorted to with an ulterior motive. Disposal of the foetus should be regulated and its commercialisation should be made an offence.

In any case, now that abortion is legal, it may not be expedient to waste such valuable biological material. Even assuming that legislation of foetal tissue research and transplantation might result in an increase in the number of abortions, this by itself as observed by one medical scientist, should not dictate an immediate halt to the technology. "Why should preventing a possible increase in a legal act be more important than saving millions of victims from suffering and early death? Only an irrational society has a greater interest in potential life than in actual life . . . ."

All said, "society cannot afford to turn its back on medical progress...By steering medical progress carefully, it can prevent abuse and preserve the promise of dramatic medical cures". With adequate safeguards and, proper balancing of the moral, ethical, social, legal and practical considerations, modern technological advancements can bring hope and happiness in the lives of innumeral hapless and dejected patients.

15

PROTECTING JOBS AND THE UNBORN CHILD*

Surrogate motherhood, commercialisation of foetal tissues and so on, are already matters of judicial and public debate. Another issue which has started causing concern is the protection of the rights of women working in places which might cause harm to the unborn child. Since an unborn child has a recognised status under the law and the claims for damages in case of deformities due to negligence of the employer while the child was *en ventre sa mere* (in the mother's womb) are well known, employers seek to safeguard their business interests by drawing up rules or policies prohibiting women capable of bearing children from taking jobs in positions which expose them to substances or situations hazardous for the foetus.

Various factors and substances are hazardous to the proper growth of the unborn child such as emotional stress, diet and exposure to certain chemicals. These may cause congenital malformation or result in abortions or other abnormalities. For instance, lead has a teratogenic effect on the central nervous system. The question that arises is whether a woman can be

---

* *Indian Express*, September 11, 1991

debarred from employment to protect the interests and health of her child-to-be?

Apart from barring discrimination on grounds of sex, our Constitution provides for protective discrimination in favour of women. These provisions are now being invoked more often because of awareness amongst women of their constitutional rights. Administrative and judicial authorities have been protective of women's interests where marriage, maternity or pregnancy become the subject of disputes. The famous air hostesses case (*Air India* v. *Nargesh Meerza*) decided by the Supreme Court in 1981 is one example. However, when it comes to protecting the life and health of an unborn child, how does one reconcile this with a woman's constitutional right to equality and the right not to be discriminated against on grounds of sex? Can employers keep women out of certain jobs to avoid potential suits for damages by the children of women working on jobs which might pose a danger to the foetus? Or, is a woman's right to employment absolute irrespective of its impact on the child-to-be? If the employer excludes a woman from such jobs, can she file a suit on grounds of discrimination based on sex? If she continues with the job knowing that its impact on the foetus will be harmful, can a child born with some deformity or infirmity sue the employer for damages?

In this context, reference may be made to American case law. In *Zuniga* v *Kleberg County Hospital* (1982) a female X-ray technician was asked to resign when she became pregnant. The plea taken was that X-ray radiation exposure could have adverse effects on the foetus which might, at a later stage, give rise to a law suit against the hospital by the child. The woman was not entitled to leave of absence, sick leave or health insurance cover.

Zuniga brought a sex discrimination action against the hospital. The district court held that "potential harm to the foetus from radiation exposure and the possibility of a tort suit by a future damaged child constitute a business necessity justifying termination". On appeal, however, the case was reversed because, according to the appeal court, the hospital "failed to use available, alternative and less discriminatory ways of attaining its business purpose."

In *Wright* v *Olin Corp.* (1982) the court held that unborn children were analogous to invitees and licencees and their safety is as important as the safety of business licencees or invitees upon an employer's premises. In another case, *Hayes v Shelby Memorial Hospital* (1984) where the services of an X-ray technician were terminated on her pregnancy, the court expanded the business necessity defence to cover foetal protection but refused to classify the foetus as a business invitee or licencee. It also held that the hospital failed to look into alternative jobs or assignments within the hospital which Hayes could perform without peril. The argument of potential litigation liability was also rejected as this would amount to shifting the focus from the interest of the foetus to self-interest of the hospital.

In a more recent case in 1989, *International Union* v *Johnson Controls*, the Supreme Court accepted that employers can bar fertile women from jobs that might endanger a foetus. Under a foetal protection policy, Johnson Controls, dealing in manufacture of batteries, sought to exclude women capable of bearing children from working in jobs where "the lead level of their blood might rise above thirty micrograms". A provision to compensate for lost earnings was also made. The court held that the business in which the Johnson's were engaged required the use of toxic substances and hence it was necessary to adopt a foetal protection policy to run the business safely. Besides, under the general industrial safety policy also, an unborn child has to be protected from substantial risk of harm resulting from exposure to toxic substances.

In the latest known case, however, the California Court of Appeal, in *Johnson Controls* v *Fair Emp. and Hous Common*, (1990) held that Johnson's foetal protection policy is "class-based exclusion rather than a neutral practice," A woman was denied employment at one of Johnson Controls' automotive battery plants because the job was subject to female exclusion under its foetal protection programme. The court held that this was in violation of the state's sex determination law and amounted to gender discrimination.

In India, there are some statutory provisions in industrial laws, under which women and children are not to be employed in hazardous positions but the focal point is not the foetus. Though most women may not themselves take up jobs which they know might pose a threat to their offspring, a balance needs to be judiciously and legally maintained between their rights and those of the unborn child. Women should not be denied jobs except where there are genuine risks of damage to the foetus despite use of techniques and devices to mitigate the baneful effects. Employers should frame foetal protection policies under which women should be excluded from certain positions and given alternative assignments.

SECTION IV

ABORTION/COMPULSORY STERILISATION

16

TO DO OR NOT TO DO*

To have children is a basic right of every woman. However, if persons afflicted with severe genetic disorders have children, not only are such offspring a burden on the society but also a source of agony to many. This poses certain questions. Should such parents be allowed to give birth to abnormal children? Or should there be compulsory sterilisation of the mentally retarded?

Positive and negative eugenics have been known since long. Selective breeding for the creation and maintenance of a superior Guardian class was the ideal of Plato in the *Republic.* Darwin suggested selective breeding techniques (*Descent of Man* and *Selection in Relation to Sex, 1871*). The idea of selective breeding is once again gaining eminence in the present day world.

Pre-natal screening for a number of genetic diseases is now made possible by medical advancements. Besides, when a woman suffers from serious and incurable mental or physical abnormalities sterilisation could be a way out. However, in cases where the woman is of a sound mind, there is not much of a legal problem. Vital legal issues arise when such a woman is not in a mental state to give free consent since it is a subject that touches upon a most delicate and personal aspect of human life—the integrity of one's body.

---

* *Hindustan Times*, February 21, 1987.

There are statutory provisions for eugenic sterilisation in many states of America. Indiana passed the first law in 1907. One of the earliest cases was *Buck* v. *Bell* (1922) where the Supreme Court upheld a Virginia statute providing for the sterilisation of inmates in state-supported institutions who were suffering from hereditary insanity and imbecility. In this case, sterilisation of a feeble-minded female inmate was in question. She was the daughter of another feeble-minded inmate and the mother of an illegitimate feeble-minded child. Justice Oliver Wendell Holmes observed:

"It is better for all the world if instead of waiting to execute degenerate offspring for crime, or to let them starve for their imbecility, society can prevent those who are manifestly unfit from continuing their kind . . . . Three generations of imbecility are enough".

In a much later case in 1962 *(In re Simpson,* Ohio, 1962) the court upheld a sterilisation operation mainly on the ground of economic interest of the state. The court stated:

"The girl is likely to become pregnant repeatedly and produce children for whom she cannot provide even the rudiments of maternal care. There is the further probability that such offspring will also be mentally deficient and become a public charge for most of their lives . . . . To permit Nora Ann to have further children would result in additional burdens upon the country and state welfare departments . . . . "

In this context two recent cases are from England and Canada. In the English case, (*In re D*, (1976)), the parents sought compulsory sterilisation of their retarded daughter, D, aged 11. The child was handicapped and mentally backward and suffered from epilepsy. Physically also she was overdeveloped for her age. The parents therefore decided that they would apply to have her sterilised when she was about 18.

She was sent to a special school and showed some signs of improvement. The mother, however, felt that she might be seduced and possibly become pregnant and give birth to an abnormal child. A doctor was consulted who also agreed that D should be sterilised.

Arrangements were made for the operation but an educational psychologist applied to the court to make D a ward of the court. She also sought an order to prevent the proposed operation from being carried out. The court refused to permit the sterilisation operation upon the child and held that the operation was "neither medically indicated nor necessary" and that it would not be in the interest of D that she should be sterilised. Such an operation, according to the court, involved a deprivation of a basic human right, viz., the right of a woman to reproduce. If such an operation is performed for non-therapeutic reasons and without her consent, it would be a violation of the woman's right.

In this case, D was not in a position to give an informed consent because of her mental condition. But at the same time evidence showed that there was an improvement in her condition and it was not unlikely that in later years she would be in a mental state to understand and appreciate the implications of the operation and make her own choice. Besides, chances of such medical advancements which would reduce the child's disabilities and retardation could also not be ruled out in this case. For all these reasons, the court refused to permit compulsory non-therapeutic sterilisation on D.

In the Canadian case (*In re Eve*, 1981) also the same issue was involved before the Supreme Court of Prince Edwad Island, Canada. Eve was a 24-year-old woman who was moderately retarded. She suffered from "extreme expressive aphasia", i.e., inability to speak and comprehend. There had been no improvement in her condition for ten years. The girl was physically attractive but was incapable of functioning as a mother and caring for a child, should she become pregnant.

The immediate cause for the mother's application seeking authority to have Eve sterilised was the fact that Eve attended a school for retarded adults during the week and returned home on weekends. She had developed a close acquaintance with another male student at the school who was also retarded though to a lesser degree than Eve. When the mother became aware of this situation she was quite concerned that Eve might possibly and innocently become pregnant. The care of the child would inevitably be left with the mother who, being a widow of sixty, found it difficult to cope with this responsibility. There was also a deep concern as to the emotional effect a pregnancy and subsequent birth might have on Eve.

The trial judge, however, dismissed the mother's application for authorisation to consent to a sterilisation operation being performed on her mentally retarded daughter. In appeal, the order was reversed and it was held that "this court should, in Eve's welfare and best interests, exercise its *parens patriae* jurisdiction and consent to the sterilisation of Eve." According to Justice Campbell, "there is no evidence before the court to suggest that she would suffer injury in the physical sense by giving birth to a child, neither is there prospect of her economic suffering . . . . The injury, if any, would be in the social or emotional context. Eve has the potential to live a happy life within a protected environment. However, the prospect of her pregnancy has, and will continue to place restrictions upon her freedom of action . . . . Without the protection of a permanent sterilisation the protected environment will become a guarded environment and the loss to Eve in terms of her social options and her relative freedom would cause substantial injury of sufficient degree . . . . "

The present case was distinguishable from the earlier English case in so far as the child in that case, did show signs of improvement. Eve, on the other hand, showed no improvement and there was no medical basis to believe that she will ever be able to overcome her disabilities.

In India, an adult of sound mind can opt for sterilisation but there has been no occasion to consider such a thing in the case of someone who is mentally incompetent. It is, however, logical that in the interest of public health, welfare of society and of the family including the parents and children born of the mentally incompetent, the state should, with proper safeguards against arbitrariness, exercise a right to restrict the bearing and begetting of children with an inherited tendency to mental feebleness and deficiency. The right to procreate must be given pre-eminence in normal situations but the protection of such right would be "a meaningless gesture for a woman who had lost her mind and was permanently condemned . . . . " and whose children were likely to be abnormal too.

17

ABORTION: WHO HAS A SAY?*

On petitions filed by two boy friends in two different courts in Canada, two pregnant women have been prevented by court orders from having abortions. Mr G., a boy friend of Ms. D., moved the court seeking an injunction that would save his baby from being aborted. The lady is, reportedly, appealing against this decision to the Supreme Court of Ontario. In the other case, Mr J. has got an injunction from the Quebec Superior Court against Ms. C that forbids her from undergoing abortion. The petitioner claimed that they were planning to have the baby while living together. Just five days after the couple split up, the lady decided to have the foetus, which was in its 18th week, aborted. Thus by these unprecedented orders the foetus have been made wards of the court. Such orders are bound to have far-reaching consequences and raise various issues of diverse nature, e.g., the women's right over her own body and right to make the choice whether or not to proceed with the pregnancy; the man's desire to have progeny; the issue of familial or marital harmony; the interest of the child if an unwanted pregnancy is thrust upon the women, and so on.

So far as abortion laws are concerned, they give a very wide degree of autonomy to the woman, subject however to certain constraints like the circumstances under which the

---

* *Hindustan Times*, July 21, 1989

pregnancy may be terminated, the length of pregnancy, the proper authorities who can perform the operation, etc. The woman does not need anybody's consent to undergo the abortion if she fulfills the other requisite conditions provided under the laws. Most of the countries permit abortions. In India, the Medical Termination of Pregnancy Act, 1971 makes liberal provisions in favour of women.

In fact abortion laws have sometimes come into conflict with family laws and given rise to problems. In *Satya* v. *Siri Ram* (Punjab & Haryana, 1983) a wife got herself aborted twice against the wishes of her husband and his family who were desirous to have a child in the family. The husband filed a petition for divorce against the wife on ground of cruelty. The same was granted. The court held that the conduct of the wife amounted to mental cruelty. Reference was made to an English case, *Forbes* v. *Forbes* (1955) where it was laid down that

"If a wife deliberately and consistently refuses to satisfy her husband's natural and legitimate desire to have children, and the deprivation reduces him to despair and affects his mental health, the wife is guilty of cruelty."

In *Satya's* case, the court also emphasised the general principle underlying the Hindu law of marriage and sonship and the importance attached by Hindus to the principle of spiritual benefit of having a son.

In yet another case decided by the Delhi High Court in 1987, *Sushil Kumar Verma* v. *Usha*, it was held that aborting the foetus in the very first pregnancy by a deliberate act without the consent of the husband amounts to cruelty entitling the husband to have a divorce. In this case, the wife who had become pregnant within one month of her marriage underwent an abortion in a government approved centre with the help of a registered medical practitioner. She did not consult her husband. His petition for divorce on the ground of cruelty was dismissed by the district court, but on appeal, the High Court granted it. This implies that the wife cannot undergo an abortion unless she seeks the husband's consent, even though the specific law on abortion, i.e., the MTPA, does not require such consent.

In this context, reference may also be made to some English cases where one of the spouses deprived or denied the other the opportunity and joy of becoming a parent. In *Fowler v. Fowler* (1952) the court made a distinction between a wife's refusal to have children and a husband's similar refusal. It observed that whereas in a man's case the refusal would amount to infliction of cruelty, in the woman's case, her refusal might be due to fear of labour, pain and suffering etc. "This is very unnatural and unfortunate" observed the court, "but it is not cruelty unless she has also an intention to inflict misery on her husband".

Courts have been quite sensitive to the sentiments of the spouse who is denied parenthood. This is understandable. But the question is, how does one reconcile this with the right of a woman to autonomy and freedom over her own self, and more so, when even the specific statutory provisions absolve her of any duty to consult her husband before undergoing an abortion. The issue is extremely sensitive and there is a need for a harmonious balance between the sentiments of both the parties. This is essential where there is a bond of matrimony. Procreation and progeny are concomitants of this bond. Denial of this may or may not be viewed as matrimonial cruelty. But what are the scales and standards of propriety and sentiments in a relationship which does not have a legal stamp of matrimony even though it may not be socially unacceptable or uncommon in a given society? There is no question of *matrimonial* cruelty here. In such cases what are the rights and duties of the partners? Are they bound by any contract? Does a breach of any promise or understanding have any sanction behind it?

18

WOMB REMOVAL: PLIGHT OF MENTALLY RETARDED*

Positive and negative eugenics has been known for centuries. Selective breeding for the creation of a superior guardian class was the ideal of Plato in *The Republic*. Darwin suggested selective breeding. Selective breeding is, in fact, gaining a lot of eminence in the present day world. The advancements in medical science and technology have made pre-natal screening for a number of genetic and other disorders, apart from determination of sex of the unborn child which has gained notoriety in our society, possible. In case of a disorder, the pregnancy may be terminated.

Another aspect of the issue is the sterilisation of a woman who suffers from serious and incurable mental or physical abnormalities. Where the woman is of sound mind and able to give consent, there is not much of a legal problem. However, vital legal and ethical issues would arise if the woman is not in a mental state to give free and informed consent, since it is a matter which touches upon a most personal and delicate aspect of human life—the integrity of one's body.

Against this backdrop, the news of womb/uterus removal of 11 mentally retarded girls in the Maharashtra Government-run institution for mentally deficient girls, near Pune, is significant.

---

* *Hindustan Times*, February 17, 1994

According to Press reports, the Director of the Department of Woman, Child and Handicapped Development had invited the doctors to perform the operations. Twenty-one women were initially selected for the operation. Four were rejected as their haemoglobin count was low and they were found unfit for the surgery. The rest were taken to the state-run Sasoon General Hospital in Pune.

On the intervention of the Chief Minister the operations were ordered to be stopped but by that time, eleven had already been performed. Many leading doctors and gynaecologists have justified the removal of uterus as being "appropriate" both from the human as well as practical point of view. They argue: "The retarded women had mental ages between two and four years. They were unable to care for their personal hygiene, particularly after they had attained puberty. Many a time they were made targets of sexual assaults and became pregnant".

The news of such operations has raised a furore and naturally so, for it does have a stirring and stinging impact on our minds and emotions. It is, however, important to view the issue objectively and practically, without succumbing to emotions, sentiments and legal technicalities.

Some countries make statutory provisions for eugenic sterilisation, for instance, in the United States, Indiana passed the first law in 1907. In one of the earliest cases, *Buck* v. *Bell* (1922), the Supreme Court upheld a Virginia statute providing for the sterilisation of inmates in state supported institutions, who were suffering from hereditary insanity and imbecility. In this case, sterilisation of a feeble minded female inmate was in question. She was the daughter of another feeble-minded inmate and the mother of an illegitimate feeble-minded child. Justice Oliver Wendell Holmes observed:

"It is better for all the world if instead of waiting to execute degenerate offspring for crime, or to let them starve for their imbecility, society can prevent those who are manifestly unfit from continuing their kind."

There have been other similar cases also. A reference may be made of a Canadian case—*In re Eve,* 1981,—before the Supreme Court of Prince Edward Island. The mother of a

24-year-old retarded girl Eve, filed an application seeking authority to have her sterilised. The mother feared that Eve, who was attending a school for retarded adults and had developed friendship with another retarded male student, might innocently become pregnant. The care of the child would obviously be left with the applicant/mother who was a sixty-year-old widow and not in a position to cope with the responsibility. There was also a deep concern as to the emotional effect a pregnancy and subsequent child birth might have on Eve. The trial judge dismissed the application, but on appeal, the order was reversed. it was held:

"This court should, in Eve's welfare and best interests, exercise it *parens patriae* jurisdiction and consent to the sterilisation."

And further, the court observed: "Eve has the potential to live a happy life within a protected environment. However, the prospect of her pregnancy has, and will, continue to place restrictions upon her freedom of action . . . . Without the protection of a permanent sterilisation the protected environment will become a guarded environment and the loss to Eve in terms of her social options and her relative freedom would cause substantial injury of sufficient degree . . . . "

Thus, at times it might be in the best interests of the girl and all concerned to resort to a sterilisation or surgery to prevent pregnancy. The girl is unlikely to be in a position to give an informed consent if she is imbecile or mentally retarded. It is therefore very important to provide for safeguards. The right to procreate must be respected and given pre-eminence in normal circumstances but the protection of such right would be "a meaningless gesture for a woman who had lost her mind and was permanently condemned . . . . " and whose children were likely to be abnormal too.

Each case, however, has to be considered on its own facts and conditions, like, the magnitude and gravity of the affliction, prospects of recovery, chances of giving birth to a normal child, best interests of the girl and the prospective child, and so on. Law and policy makers and enforcement authorities cannot overlook the fact that sterilisation of an incompetent is a very drastic measure to slaughter a woman's most sublime, cherished and intimate potential of child-bearing. It should not therefore be resorted to except in the rarest of rare cases where the scales of compassion and expediency demand a tilt in favour of such action.

SECTION V

DAMAGES FOR WRONGFUL BIRTHS

19

FAILURE OF FP DEVICES: AWARD OF DAMAGES*

Family planning devices are being resorted to by innumerable couples all over the world. However, when despite submission to these devices a woman conceives, the dilemma of the couple is understandable.

The question is, what is the relief they are entitled to? This issue has become a subject matter of judicial determination during the last few years.

There has been a case on the point even in India in 1983. A mother had undergone an abortion at an Ahmedabad Municipal Corporation hospital but notwithstanding that, she delivered a child. The mother filed a suit for damages in the city civil court. The judge accepted her plea and ordered the Municipal Corporation to pay her Rs. 22,100 by way of damages.

This was a case where despite an abortion a child was born. There have been cases where a husband or wife had resorted to surgery by way of vasectomy or sterilisation and still the wife conceived. An English case, *Udale* v. *Bloomsbury Area Health Authority* (1983) is an interesting case on the point where a woman had undergone sterilisation after having a number of children. The operation failed because of the doctor's negligence and a healthy child was born. The health

---

* *Hindustan Times*, September 4, 1985

authorities admitted their liability for negligence but there was some dispute regarding the quantum of damages which were claimed.

The defendants agreed that the plaintiff might claim damages for the unsuccessful or ineffective sterilisation operation, the anxiety of an unwanted pregnancy and for the pain and suffering she underwent for the subsequent sterilisation. Damages were also agreed on account of her loss of earnings, pain and suffering during the pregnancy but the court refused to award anything towards the cost of building another bedroom for the additional child or the cost of its upbringing. A very important aspect of such claims was duly considered by the court. In the words of the judge: "It is highly undesirable that any child should learn that a court has publicly declared his life or birth to be a mistake — a disaster even — and that he or she is unwanted or rejected. Such pronouncements would disrupt families and weaken the structure of society". The court granted £ 8,000 by way of general damages for the pain and suffering and £ 1025.20 for loss of earnings.

In an American case *Mason* v. *Western Pennsylvania Hospital* (1982) a mother who gave birth to a child despite a sterilisation operation was given costs associated with pregnancy and delivery but no damages for financial or emotional costs for raising the child were given for reasons of public policy.

Very recently there have been two interesting cases from English courts — one in which a healthy child was born and the other where a child with congenital defects was born despite a sterilisation operation.

In *Thake* v. *Maurice* (1985) after a vasectomy operation on the husband, he became fertile and the wife conceived and gave birth to a healthy child. The court awarded costs for the birth of the child and her upkeep but not for the distress, pain and suffering undergone by the plaintiffs. An amount of £ 9,677 was granted.

In the other case, also decided this year, *Emeli* v. *Kensington,* the plaintiff who had three normal children underwent a sterilisation operation but conceived. She discovered that she was 20 weeks pregnant and refused to have an abortion. A child with congenital abnormalities was born. The mother claimed damages against the defendant hospital for the pregnancy, and the birth and upkeep of the child. It was argued for the defendant that the plaintiff's conduct in refusing to undergo an operation was unreasonable so as to absolve them of their liability and also that it was contrary to public policy to award damages in such cases. The court did not accept these contentions and held that the operation for sterilisation was performed negligently. As the child needed to be under constant supervision, both medically and by her mother and also had speech problems requiring therapy, an award of £ 3,000 for pain and suffering till the time of trial and £ 10,000 for the future loss of amenity and pain and suffering which the child has to undergo for many years, was made.

With more and more couples resorting to contraceptives, abortions and operations for restricting their families and avoiding unwanted children, cases of conceptions and births despite safeguards cannot be ruled out as is evident from the cases discussed. With the people's awareness of such claims being decreed by courts, the number of cases are also bound to come up.

Clear guidelines or principles must be evolved about the circumstances in which damages are to be awarded and the amount of the damages. Also, it is not improbable that unscrupulous people may exploit the situation—especially in cases where birth of a child is attributed to failure of a contraceptive device. The facts are too personal to be verified in such cases. The manufacturers, of course, can absolve themselves by a non-liability clause in case of failure but legislative and judicial clarity is needed.

20

IS COMPENSATION FOR 'WRONGFUL BIRTH' MORALLY RIGHT?*

A woman in Padra Taluka, a town in Baroda, underwent a tubectomy. Despite the operation, she delivered a child. She has now filed a compensation suit against the Gujarat health minister and the doctor, demanding Rs. 2 lakh for the child.

Cases such as these are no longer a rarity. They drag the medical profession, manufacturers of contraceptive devices and pharmaceutical companies to court and often creating emotional insecurities in children.

A brief survey of cases will reveal how often such actions are being brought to courts and how the courts seem "lost in the quagmire of ideological and moral issues" involved in such claims.

In 1983, there was such a case decided by a court in Ahmedabad. A woman underwent an abortion at a local municipal corporation hospital. Notwithstanding the operation, she delivered a child. She then filed a suit in the city civil court. The judge accepted her plea and ordered the municipal corporation to pay her Rs. 22,100 by way of damages.

Amongst the earliest English cases, reference may be made to *Sciuriaga* v. *Powell* (1980). The plaintiff, a single woman, sought termination of her pregnancy by the defendant doctor for a fee of £ 150. The same was performed negligently and

* Hindustan Times, September 29, 1989

she delivered a healthy child. One of the arguments of the defendant was that "it was repugnant to the peoples' sensibilities that damages should be awarded for a failed termination of pregnancy". The judge at the Queen's Bench division granted the plaintiff £ 18,750 by way of damages for loss of earnings up to the date of the trial, future loss of earning capacity, impairment of her marriage prospects and for suffering. In appeal, the award was reduced by £ 4,250. The court ignored policy considerations, but observed: "In doing so, it must not be taken as assenting to the view that they would be irrelevant in every case . . . ."

On grounds of public policy, an American court in *Mason* vs. *Western Pennsylvania Hospital* (1982) declined to give damages for financial or emotional costs for raising a child born after an unsuccessful sterilisation, though the cost associated with pregnancy and delivery were decreed.

Public policy considerations, however, were ignored in *Emeli* v. *Kensington Area Health Authority* (1984) where a woman gave birth to a handicapped child after an unsuccessful operation.

The emotional or psychological damage caused to the child, if it comes to know that its birth has been an unwelcome event warranting its parents to go to the court to claim damages should not be ignored. The American courts in some of its earlier cases, e.g., *Custodio* vs. *Bauer* (1967) and *Sharlock* v. *Stillwater Clinic* (1977) rejected the "emotional bastard" or damage argument but in later cases, the courts were sensitive to this aspect.

The Arkanasas Supreme Court, in *Wilbur* v. *Kerr* (1982) described as significant the damage which could result to the child later learning that it was unwanted to the extent that its parents "went to court to force someone else to pay for its raising."

It may be pointed out that a claim for recovery of damages for "wrongful" birth is shocking to many. As observed by Justice Ognall in *Jones* v. *Berkshire Health Authority* (1986), "speaking purely personally" (he is surprised) "that the law acknowledges an entitlement in a mother to claim damages for the blessing of a healthy child" (and that) "those who are afflicted with a handicapped child or who long desperately to have a child at all and are denied that good fortune would regard an award for this sort of contingency with a measure of astonishment".

Thus, we find that the reactions of the public and the judiciary to this sensitive issue are bound to vary. Even the courts have no fixed scales or criteria to arrive at the quantity of damages paid in such case, even if on principle they decide that some damage has been caused.

Some judges view public policy considerations and set off pain and distress by the joy of birth as important whereas others do not feel so.

In view of the increase in the number of couples resorting to family planning operations or the use of contraceptives, and the widespread legal and moral implications of wrongful conception/birth claims, it is important to evolve clear guidelines and principles. What should be the extent of the doctors/defendants' liability?

If too liberally construed, such suits would impose a lot of stress and strain on the medical profession and would also open the floodgates of litigation. Unscrupulous persons, under guise of failure of contraceptive devices, might try to make a bargain out of the birth of a child.

Also, the emotional impact of such a litigation on the child needs to be carefully considered. In short, a harmonious balance between the interests of the child, the medical profession and the right of the parties to plan their families according to their needs and circumstances is needed.

As aptly observed in a recent case from New Jersey (*Schroceder* v. *Perkel*, 1981): "The problems of wrongful conception and wrongful birth involve an evaluation not only of law, but also of morals, medicine and society. Thus, it is not surprising that the same issues may elicit divergent judicial response." No wonder then that "a public debate not a judicial one should determine the standards of medical care and liability"

21

THE BURDEN OF AN UNWANTED CHILD*

Can the birth of an unwanted child ever be a matter of court litigation for a claim for damages? This is an issue that has come up for judicial pronouncement several times in other countries and very recently, the awareness that birth of an unwanted child could be a fit cause of action seems to have dawned upon us too.

Use of family planning measures by way of contraceptives and other procedures—surgical, injective and medicinal, is undoubtedly on the increase. People seek to restrict the size of their families or prevent the birth of a child due to social, economic, medical or eugenic reasons. In countries which face the problems of poverty coupled with population explosion, there are strong measures and numerous appeals to use contraceptives and or go in for tubectomy or vasectomy.

Incentives and disincentives are offered to achieve the goal of population control. Legalisation of abortion is one measure—though obliquely—to help people restrict family size. A pertinent question in this context is, can a mother who delivers a child despite having undergone an abortion or tubectomy claim damages for the birth of such a child? The courts seem to be "lost in the quagmire of ideological and moral issues" involved in such claims. Early cases denied any damages for unsuccessful sterilisation resulting in a normal, healthy unwanted baby — for example, in *Shaheen* vs *Knight* (1957) a

* *Pioneer*, March 29, 1995

case in the UK, the husband had undergone vasectomy to limit the size of his family. Despite the operation, the wife gave birth to a fifth child. He sued the physician and sought recovery for "the additional expenses of supporting, educating and maintaining the said child, until maturity . . . ."

Denying the same, the court observed: "We are of the opinion that to allow damages for the normal birth of a normal child is foreign to the universal public sentiment of the people. To allow damages in a suit such as this would mean that the physician would have to pay for the fun, joy and affection which plaintiff Shaheen will have in the rearing and educating this child. Many people would be willing to support this child were they given the right of custody and adoption, but according to the plaintiff's statement, he does not want such. He wants to have the child and wants the doctor to support it. In our opinion, to allow such damages would be against public policy."

In *Doiron* v. *Orr* (Canada, 1978), the judge stated that he would have been prepared to award damages for mental anguish caused to the plaintiff but was adamant in his refusal to accept that in such a case there could be liability for the cost of bringing up an unwanted child. He said, "I find this approach to a matter of this kind which deals with human life, the happiness of the child, the effect upon its thinking, upon its mind when it realised that there has been a case of this kind, that it is an unwanted mistake and that its rearing is being paid for by someone other than its parents, is just simply grotesque."

In a UK case in 1980 (*Scuriaga* v. *Powell*), a single woman delivered a healthy child by a caesarian section despite an operation for termination of pregnancy. The judge at the Queens Bench Division granted damages for loss of earnings, future loss of earning capacity, impairment of her marriage prospects and for suffering.

The American courts in *Mason* v. *Western Pennsylvania Hospital* (1982) refused damages for financial or emotional costs for raising a child born after an unsuccessful sterilisation. The validity of wrongful pregnancy action was upheld in *Udale* v. *Bloomsbury Area Health Authority* (England, 1983) in which damages were given for pain and suffering alongwith loss of earnings following a negligently performed operation.

The court, however, rejected an award in respect of the cost of bringing up the child. Justice Jupp, in his judgement, reiterated that the joy of having the child and the benefits it brought in terms of love, should be set off against the inconvenience of financial disadvantages resulting from its birth. He observed "It is an assumption of our culture that the coming of a child into the world is an occasion for rejoicing."

On the other hand, in *Benarr* v. *Kettering Health Authority*, a British court in 1988 went to the extent of awarding damages even to cover the private education of the child. In this backdrop, a reference may be made of the Indian position. Though there have not been many reported cases on the point, the issue seems to have caught the attention of some people.

In 1983, a woman who had undergone abortion at a local municipal corporation hospital in Ahmedabad delivered a child notwithstanding the operation. She filed a suit in the city civil court for damages for the birth of the unwanted child. The court ordered the municipal corporation to pay her Rs. 22,100 by way of damages.

A few years later, a woman in a town in Baroda delivered a child despite a tubectomy. She filed a compensation suit against the Gujarat health minister and the doctor, claiming Rs. 2 lakh as compensation for the birth of the unwanted child. The outcome of this litigation, however is not known.

And now, just a few days ago, as reported in the newspapers (January 7, 1995), a woman who already had three sons, gave birth to a son in a village in Patiala (Punjab) despite a tubectomy operation performed in a Government civil hospital. The case has been taken up with the health department of the Punjab government by the general secretary, Punjab State Youth Wing of the Janata Dal.

The issue arising out of such claims are complex. It is obvious that a patient undergoing sterilisation/abortion may not be wanting any more children and one strong, or probably the only, reason for this — in our country especially — could be economic. The birth of an additional unwanted child undoubtedly means a heavy financial liability on the parents.

As aptly pointed out by an English Court in *Thake* v. *Maurice* (1984) "every baby has a belly to be filled and a body to be clothed", more so since we have no provisions of state subsidy or social security, the entire burden of rearing the child rests on parents who may have no means to do so.

The ramifications of unwanted pregnancy and child birth would vary depending on whether it is a healthy child or an abnormal/unhealthy child; and given our peculiar social cultural conditions, even the gender of the child is bound to make a lot of difference. While the parents may accept or compromise where it is a healthy male child, the "burden" could be enormous in the case of a girl or an abnormal baby. This aspect cannot be entirely ignored. On the other hand, policy and ethical considerations in paying damages for the birth of a child cannot be overlooked either. The emotional affect on the development of the child and on parent-child relationships could be grave. Each case, however, would depend on its own peculiar facts but a balance has to be maintained between the practical difficulties of the parents in bringing up that child and the impact of such claims on the psychology of the child. Also, the risk of influx of such claims needs to be guarded against.

SECTION VI

SEXUAL IDENTITY ABERRATIONS/ CONFLICTS AND TRANS-SEXUALISM

22

CHANGE OF SEX BY CHOICE: LEGAL IMPLICATIONS*

The advances made in medicine are truly miraculous. Until a few decades ago it was difficult to imagine that a child could be born in a test tube, that a man could be changed into a woman and a woman into a man. These achievements of science however have given rise to many social and legal issues.

Ever since 1952 when Christine Jorgenson's biological sex was changed by surgery, the world has been baffled by such questions as — what is the status of a person who has changed his sex? Is he to be treated as a female or a male? What about birth certificates and passports, do they have to be changed; what are the implications of a sex change in matters of marriage, inheritance an social security?

A trans-sexual is a person, anatomically of one sex who firmly believes he belongs to the other sex. This belief is so strong that he is obsessed with the desire to have his body, appearance and social status altered to conform to that of his "rightful" gender (Douglas K. Smith in "Trans-sexualism, sex-reassignment surgery and the Law" 56 *Cornell Law Review*, 963 (1971).

It is a psychological state the clinical term for which is "gender dysphoria syndrome." Such a condition is normally

---

* *Hindustan Times*, November 18, 1979

attributed to a hormonal imbalance. The suffering of such people, it is believed, is "beyond belief" and the only treatment is surgery, i.e., sex reversal or change.

Surgical sex change is legal in some countries like Switzerland, Canada, Belgium and Great Britain if performed after thorough medical evaluation for therapeutic reasons. Argentina distinguishes between surgery done for a change of sex and that performed for sex clarification. The former makes the surgeon liable whereas the latter is justifiable, but in Germany, it is not lawful. In 1969 a boy who had undergone sex reassignment surgery wanted his birth certificate to be changed accordingly. It was held by the court that the birth certificate that correctly stated the boy's sex at birth could not be altered. The court, however, did not rule out the possibility that such petitions might be granted in the future if medical advances could show that there had been more than a change in appearance.

This indicates the impact medical advances are likely to have on law and judicial pronouncements. Some American States expressly provide for birth-record changes for those who have undergone sex reassignment surgery.

There are about a dozen such places in United States where sex-change operations are undertaken. The patient is first interviewed, then extensively tested by a team consisting of a psychologist, a psychiatrist and a surgeon. When such tests indicate that the person is physically and mentally a true trans-sexual and needs treatment, hormonal therapy is undertaken to alter the secondary sex characteristics. Further tests of a psychological nature are then conducted to assess the patient's reaction. Surgery is performed only after the clinical staff are satisfied that he is physically and psychologically ready for a change of sex. There is a post-operative follow up too which facilitates his assimilation into society in the new sex role.

With the passage of time and as medical science advances further, cases of sex change are bound to increase and pose problems of varied kinds.

The first court to hold a trans-sexual marriage as valid was an American Court in *M.T.* v. *J.T.*, in 1976. This was a marriage between two biological males. The court ordered the husband to pay spousal support to the wife since this male spouse had become "physically, psychologically and anatomically female."

In an earlier English case, *Corbett v. Corbett* (2 All E.R. 33 1970), however, such a marriage was annulled. George Jamieson, the 'wife' was born and registered at birth as a male and brought up as a male. At the age of 25 he underwent sex reassignment surgery and adopted a different name, April Ashley. Arthur Corbett, April's husband was a transvestite i.e. liked to live and dress up like one of the opposite sex. He had also indulged in sexual relations with numerous men. Both parties sought annulment of the marriage on grounds of incapacity or wilful refusal to consummate the marriage. The husband further alleged that the marriage was between two males and therefore void.

The judge reasoned that since marriage is essentially a relationship between man and woman the validity of the marriage in this case would depend upon whether the respondent is, or is not, a woman. And concluded, "It is common ground between all the medical witnesses that the biological sexual constitution of an individual is fixed at birth (at the latest), and cannot be changed, either by the natural development of organs of the opposite sex, or by medical or surgical means. The respondent's operation, therefore, cannot affect her true sex." The marrige was consequently declared a nullity.

It is suggested that sex change surgical operations should be legal only in cases of bonafide medical problems requiring such treatment. In other words, the purpose should be therapeutic and not cosmetic. Some feel that a distinction should be drawn between the hermaphrodite or inter-sex and the transvestite. Whereas in the former, the sex needs clarification, in the latter case the person concerned is anatomically normal and can be clearly classified as male or female, and therefore, according to some, surgery should be legal only in the former case and not in the latter.

Others however, criticise this distinction and hold that "if physical, incongruities such as those present in a hermaphrodite are sufficient to justify corrective surgery, it should be the same in the psychological inconguity of a trans-sexual." Each case however, has to be judged on its own merits, but one thing is clear, a trans-sexual is in need of understanding both before surgery and after, at least until he has been absorbed and assimilated into society in his new capacity.

23

LEGAL IMPLICATIONS OF SEX CHANGE SURGERY*

Introduction

The miraculous advances made in the field of medicine have given rise to numerous legal issues. The one under discussion here is the legal implications of sex reassignment surgery and transsexualism.

The word "transsexualism"[1] has been coined by an American physician, Harry Benjamin.[2] A transsexual is a person who belongs to a particular sex anatomically but is psychologically obsessed with a desire to change to the other sex which, according to him, is his rightful gender. He cannot reconcile to the sex he is born in and, as pointed out by J. Morris, "[n]o true transsexual has yet been persuaded, bullied, drugged, analyzed, shamed, ridiculed or electrically shocked into an acceptance of his physique."[3] It is believed that so long as such persons live with their biological organs, they feel "miserable, morbidly longing for the 'conversion' surgery they think will make them look as they really should."[4] A transsexual finds it extremely difficult to adapt himself to the society. He

* 25 Journal of Indian Law Institute, 73–89 (1983)

is ashamed of, and embarrassed with, his features. Sometimes such a person may even develop suicidal tendencies.[5]

A transsexual differs from a transvestite and a hermaphrodite. The former merely has a desire for, and is gratified by, dressing up in women's clothes. He has no urge to physically change his sex. The latter is a person whose gender identity is not clear. He has traits of both the sexes and his problem is one of bringing his body by surgical alteration into harmony with his true sex, *i.e.*, the one which predominates in his anatomy. The transsexual, on the other hand, is a person whose biological sexual development is normal, but who strongly believes that he or she is a member of the opposite sex.[6]

Though transsexualism is a comparatively recent classification in the world of medicine, the fact is that we do find a few personalities in history who are believed to have been transsexuals.[7] The presence of transsexuals and transvestites in classical Greece is also suggested by a picture of Hercules, dressed in feminine clothes, serving his mistress Omphale.[8] However, until 1952, the phenomenon of sex change did not receive much publicity. It was in that year that Christine Jorgensen, a young man of 24, went to Denmark and successfully underwent sex change surgery after two years of observation and hormonal treatment. Danish physicians thereafter received numerous requests for such a surgery, but the authorities refused permission for performing the operation on males who were not Danish citizens of long standing. Ever since this case, however, the world has been baffled by legal issues arising in situations where sex is a material factor for determining the legal status.

Sex Determination: Relevant Criteria

There is no single criterion by which sex could be differentiated and human beings divided into male and female sex. According to Moore, "sex is not a well-defined entity, but, instead, gradations from a typical female to a typical male exist."[9] Maranon says that "there is scarcely any human being whose sex is not tainted by doubt or at least shadow of doubt."[10] According to still another view, the pure man and the pure woman are "extreme cases . . . scarcely to be met."[11]

However, according to the medical profession, there are certain general factors on the basis of which a person's sex could be determined. According to Bartholomew, there are four distinct levels at which sexual differentiation may be considered, *viz.*, genetic sex determined by the chromosome constitution; gonadic sex determined by the type of gonads; genital sex determined by the genital structure; and apparent sex determined by the secondary sex characteristics exhibited by the individual.[12]

According to Moore, the criteria to be applied for determining sex are sex chromosome constitution, gonadal sex, sex hormone pattern, internal sex organs other than gonads, genitalia, secondary sex characteristics, sex of rearing, assumed sex role and psychological sex.[13]

None of the above criteria, however, can be said to be the sole factor controlling the sexuality of a person. Taking, for instance, the genetic factor, some persons have an abnormal genetic constitution making it difficult to classify them. Besides it is the genital rather than the genetic sex which is responsible for the social manifestation of reality. Genetic sex is too abstract and is not suitable for the purposes of law, which is primarily concerned with social relations.[14] The gonadic criterion is also open to objection. A person may have the gonads of both the sexes (hermaphrodite) or may lack the gonads of either sex. As to hormones, they can be changed by drugs and the hormonal balance can be affected by castration. Then, as regards internal sex organs and external genitals, abnormalities are often met with. Secondary sex characteristics may not always be reliable. The sex of rearing conforms to social reality since the individual has been trained in that sex role, has adapted himself to that sex role and has been likewise accepted by the society as a member of that sex. But, while this criterion may be very material in determining the sex of a hermaphrodite, in the case of a transsexual it is not so because the latter is normally brought up and reared in his or her anatomical sex.[15] The factor of psychological sex, *i.e.*, the sex that an individual assumes himself to belong to, is an extremely important factor so far as transsexuals are concerned. As observed by Smith:

> The psychological sex is appealing because it is at once practical, realistic and humane . . . . For transsexuals a psychological test legally recognizes what the individual has thought himself to be all along, a conception to which his anatomy now conforms.[16]

Effect of Surgery

The sex change operation involves the removal or construction of external genital organs. For male to female surgery, the external male organs are removed and artificial female organs constructed by plastic surgery, and vice versa in female to male surgery. The process is supplemented by hormonal treatment which brings about a change in secondary sex characteristics also. The person undergoing the operation, though capable of performing sex, becomes sterile. According to J. Randell, post-operative transsexuals "will remain castrated males [or females] and they will not be made into [biological] women [or men]."[17]

Thus, biologically a person's chromosomal or genital sex remains unchanged. However, so far as social reality is concerned, the person, in his changed role, has all the apparent traits and looks of the new sex. He has left his old form to which he can never revert. It is, therefore, proper that after the surgical reassignment the social criterion for sex determination should be recognised. In practice, the state of affairs is not very happy as societal acceptability is rather low and it cannot be said that after the operation "roses now suddenly bloom . . . and pizzas are always delivered hot."[18] In fact more often than not, transsexuals suffer a set-back after the operation.[19]

Legality of the Surgical Operation for Sex Change

Amongst the legal issues that revolve around the subject, the first relates to the permissibility of the operation for change of sex. Its legality is in most cases to be decided with reference to the provision in the general criminal law applicable to consent to surgical treatment. In some countries, certain conditions have to be fulfilled before the operation can be performed while in some others, the matter may be simpler. For example in Argentina, such surgery is permissible with patient's consent if physical abnormalities are present. The law in Belgium requires that the patient must give his consent for the surgery and also there must be adequate pre-operative medical evaluation. Under the Canadian law such surgery is permissible under section 45 of the Canadian Criminal Code 1970 which reads as under:

> Every one is protected from criminal responsibility for performing a surgical operation upon any person for the benefit of that person if
>
> (a) the operation is performed with reasonable care and skill, and
>
> (b) it is reasonable to perform the operation, having regard to the state of health of the person at the time the operation is performed and to all the circumstances of the case.

In Great Britain, such a surgery would be permissible if undertaken for genuine therapeutic purposes. At the Charing Cross Hospital in London, patients are required to live in the new sex role for at least six months prior to surgery to ensure that they are markedly better adjusted in the role they desire than in the role they want to leave.

In the United States and Canada, there are about a dozen research and treatment centres where the problems of transsexuals are studied. The individual is first extensively tested and interviewed by a team consisting of a psychologist, a psychiatrist and a surgeon. It is only when the person is found to be a true transsexual and physically and mentally suited to, or needing, the treatment that hormonal therapy is undertaken. This is done to alter his secondary sex characteristics. After this, further psychological tests are undertaken to test his reaction to the therapy. The surgery follows only when the clinical staff is convinced that the person is mentally and physically ready for the complete transformation. Post-operative follow up is equally important, being aimed at helping the person to assimilate into the new sex role.

In India, the matter has not arisen before the courts so far but it is not unlikely that, before long, our courts as well as the medial profession might be faced with issues of the nature mentioned above. According to newspaper reports,[20] in 1979 an Indian girl aged about nineteen years was refused admission to a junior college at Karimganj in Bihar unless she produced a medical certificate to confirm her sex. The girl had passed her higher secondary examination in 1978 and was an athlete. Later she developed masculine features like moustache. She was also denied an opportunity to participate in a *kabaddi* tournament, a game in which she had won laurels in the past. Her doctor was of the opinion that she was undergoing sex transformation and suggested surgery to complete the process.

Strictly speaking, a surgeon called upon to perform such a surgery with the patient's consent would be well protected against criminal liability under section 88 of the Indian Penal Code, provided the operation is performed for the patient's benefit and in good faith. This section provides:

Nothing which is not intended to cause death, is an offence by reason of any harm which it may cause or be intended by the doer to cause, or be known by the doer to be likely to cause, to any person for whose benefit it is done in good faith, and who has given a consent, whether express or implied, to suffer that harm, or to take the risk of that harm.

Thus it appears that so far as criminal liability of any party is concerned the law is quite clear. The question, however, is how far would the law go to protect the civil status of a person who has undergone the surgery, for instance, in regard to employment, educational facilities, change of name and sex in documents of identity, matrimonial relationships and so on. The position in some countries on the above aspects has been discussed in the pages that follow.

Sex Change and Alterations in Documents of Identity

Change of sex raises the important question as to how far it should be reflected in birth certificates and other documents of personal identity. The legal position in this respect differs from country to country. In some countries the legislature has had to intervene. Elsewhere the position has to be gathered from case law.

In West Germany, the first such surgical operation was performed in 1931. But the courts do not seem to have accepted such operations as altering the legal status of the individual. In Berlin and Frankfurt, the courts have held that alteration of sex by surgical removal of the sexual organs will not be accepted as a basis for altering a passport or for avoiding a possible prosecution for homosexuality.[21] In a German case of the mid-sixties relating to male to female sex reassignment surgery, an application for recording a change of sex in the birth certificate was refused on the ground of want of adequate data in medical science. The court, however, did not rule out the possibility that such petitions might be granted in future if medical evidence could show that there had been more than a mere change in physical appearance.[22]

A Scottish case illustrates how the statutory provision for amendment of such records for rectifying incorrect entries could deny a request for change in birth records consequential on change of sex. The petition in that case was for amendment of the birth register under section 63 of the Registration of Births, Deaths and Marriages (Scotland) Act 1854. A married man, father of two, had developed feminine characteristics which caused breakdown of marriage. He was quite happy to act like a woman. His application for recording the change was, however, rejected. The section was construed as allowing correction only of entries originally made incorrectly and not of entries which became incorrect because of alleged changes in sex subsequent to registry at birth. Nor could the relief be granted apart from statute because the sheriff, adopting a chromosomal test of sex, held that the changes had not yet reached "the deepest level of sex determination . . . . While X could be described as an abnormal male, it would not be possible to describe him as a female."[23]

South Africa, apparently recognising the need for specific statutory provision regarding change of records, has provided by statute as under:

> The Secretary [of the Interior] may, on the recommendation of the Secretary for Health alter, in the birth register of any person who has undergone a change of sex, the description of the sex of such person and may for this purpose call for such medical reports and institute such investigation as he may deem necessary.[24]

Switzerland seems to have evolved a liberal approach even without the help of statute. As early as 1945, it was held in a case that after sex assignment surgery, the changed sex should be legally accepted. "In granting him the civic status of woman we are satisfying the most profound desire of his being while consolidating his psychic and moral equilibrium; at the same time we are facilitating his social adaptation by permitting him to lead a more normal type of life than heretofore."[25]

It is important to note the provisions of the European Convention on Human Rights guaranteeing the right to private and family life and the right to marry and set up a family. In *Van Oosterwijck* v. *Belgium*,[26] the state refused to give official recognition to the new sex of a transsexual who, born as a female, had a sex change surgery and became a male. The European Court of Human Rights held that his claim could not be determined on its merits because he had failed to exhaust domestic remedies. The European Commission of Human Rights, however, considered the case on merits. As regards the contention of violation of article 8 of the convention, which guarantees the right to respect for private and family life, it observed:

> It would appear scarcely compatible with the obligation to respect private life to force a person who on the recommendation of his doctor and by undergoing a lawful treatment has taken on the appearance, and, to a large extent, the characteristics of the sex opposite that which appears on his birth certificate to carry identity documents which are manifestly incompatible with his appearance. In such a case he would in fact be exposed to having to reveal to anyone information relating to his private life and subsequently to being excluded from certain employment, activities and relationships on account of the explanations

about his position which he had improperly been required to give.[27]

The commission was unanimous in its finding that the state had violated article 8 of the convention. A majority of the commission also held that article 12 of the convention, which guarantees the right to marry and set up a family, was also violated. In the circumstances of the case, it was "inconceivable for psychological, physical and social reasons" for the applicant to marry a man but since the state did not recognise him as a man, he was denied the right to marry.[28]

Cases and statutes in the United States offer extremely interesting material on the question of changes in birth certificates, both because the material shows the variety of approaches that might come to be adopted and because it reveals a real endeavour to grapple with the issue. Mention may be made, at the outset, of a New York case.[29] A transsexual, who had become a female by surgery, requested the Director of Bureau of Records and Statistics of the New York City Department of Health to issue a fresh birth certificate. The Board of Health of the City of New York consulted the New York Academy of Medicine on the subject, and the academy, in its turn, constituted a committee consisting of medical specialists and a lawyer. It recommended that a change in birth certificate should not be permitted in such cases. It held as follows:

> 1. [M]ale-to-female transsexuals are still chromosomally males, while ostensibly females;
>
> 2. It is questionable whether laws and records such as the birth certificate should be changed and thereby used as a means to help psychologically ill persons in their social adaptation.
>
> The committee is therefore opposed to a change of the sex on birth certificates in transsexualism.
>
> . . . The desire of conealment of a change of sex by the transsexual is outweighted by the public interest for protection against fraud.[30]

The application was denied by the board following the above recommendation of the committee. The petitioner then brought a suit for a court order requiring the agency to change her sex and issue a new certificate, but the suit was dismissed. Viewing the problem as one of administrative law, it was held that the board had not acted in an "arbitrary, capricious or otherwise illegal manner."[31]

In a later case, Judge Pecora severely criticized the report of the committee filed in the earlier case. According to him, a male transsexual who submits to sex reassignment surgery is anatomically and psychologically a female. He noted that the applicant before him who had applied for a change in the birth certificate "dresses, acts and comports himself" as a female. But for the fact that the applicant's background was known to the court, the court would have found it impossible to distinguish the applicant from any other female. On the aspect of fraud emphasized in the recommendations of the committee, he pointed out that the probability of so called fraud, if any, existed to a much greater extent where the birth certificate is allowed, without annotations of any fact, to classify an individual as a male when the individual in fact comports himself as a female. Accordingly, the judge permitted the applicant to change her name and directed the board of health to attach a copy of the order of the court to the applicant's birth certificate.[32]

Apart from these cases, Karl M. Bowman and Bernice Engle have collected useful data from transsexuals regarding change of name, passport or certificate. Four interesting cases have been mentioned by them. In the first case from California, a male who had undergone sex change operation in Mexico petitioned for change of name and civil status, but was refused relief. In the second case, also from California, a similar petition was granted, but in that case the individual's birth certificate had been made out simply in the name of "Baby S" without any specific name. In the third case, only change of name was allowed, without reference to the person's sex. In the fourth case, the individual was able to obtain a passport in the changed female sex.[33]

Controversies on this subject have been sought to be avoided by enacting suitable legislative provision in Illinois and Louisiana of the United States.[34] The Illinois law requires simply an affidavit by a physician that "he has performed an operation on a person, and that by reason of the operation the sex designation on such person's birth record should be changed." After the new certificate has been issued, the original is sealed and available only upon court order. The Louisiana statute provides as under:

> Any person born in Louisiana who, after having been diagnosed as a transsexual or as a pseudo hermaphrodite, has sustained sex reassignment or corrective surgery which has changed the anatomical structure of the sex of the individual to that of a sex other than that which appears on the original birth certificate of the individual, may petition a court of competent jurisdiction . . . to obtain a new certificate of birth.

Sex Change and its Effect on Marriage

Pre-marital Sex Change

An individual may undergo sex change surgery either before marriage or after the marriage and in both situations controversy is likely to arise. On pre-marital change of sex, the most publicised case is *Corbett* v. *Corbett*,[35] an English case. This related to the validity of a marriage. The wife, respondent in the case, was born, registered and brought up as a boy under the name of Geroge Jamieson. In 1960, at the age of 25, the respondent underwent a sex change surgery and called "herself" as April Ashley. Six months later she met the petitioner, a married man having children. Soon after the marriage the parties sought annulment of the marriage on the ground of incapacity and willful refusal to consummate the same. The husband further alleged that the marriage was between two males and was therefore void. Counsel for the wife contended that the wife should be regarded for all purposes as a woman since the law recognizes only two sexes — male and female, and in this case it could only be the female sex. He further submitted that assignment of an individual to one sex was a matter for the individual and the individual's doctor, and the law ought to accept it as determining the individual's sex. Further, it was illogical to treat the respondent as a woman for some social purposes such as national insurance and her being nursed in a female ward in the hospital and yet not to regard her as a woman for the purpose of marriage. The court, however, rejected all these contentions. In its view, validity of the marriage depended on the respondent being a woman. As all medical witnesses agreed that the biological sex constitution of an individual is fixed at birth, at the latest, and cannot be changed either by the natural development of organs of the opposite sex or by medical or surgical means, the operation cannot change the respondent's true sex, which undoubtedly was male at the time of birth. The court granted the petitioner's application for declaration that the marriage was null and void being a marriage between two biological males. According to Ormrod J.:

> Having regard to the essentially heterosexual character of the relationship which is called marriage, the criteria must, in my judgment be biological, for even the most extreme degree of transsexualism in a male or the most severe hormonal imbalance which can exist in a person with male chromosomes, male gonads and male genitalia cannot reproduce a person who is naturally capable of performing the essential role of a woman in marriage. In other words, the law should adopt, in the first place, the first three of the doctors' criteria, *i.e.* the chromosomal, gonadal and genital tests, and, if all three are congruent, determine the sex for the purpose of marriage accordingly, and ignore any operative intervention. The real difficulties, of course, will occur if these three criteria are not congruent. This question does not arise in the present case and I must not anticipate, but it would seem to me to follow from what I have said that greater weight would probably be given to the genital criteria than to the other two. This problem and, in particular, the question of the effect of surgical operations in such cases of physical intersex, must be left until it comes for decision. My conclusion, therefore, is that the respondent is not a woman for the purposes of marriage but is a biological male and has been so since birth.[36]

The judgment has not escaped criticism as will be seen from some of the points later made in this article.

Two South African cases also involved the question of validity of the marriage of transsexuals. In both of them one of the parties had undergone a sex reassignment surgery prior to the marriage. The first case is of 1970.[37] The husband petitioned for conjugal rights against his wife who, prior to the sex change operation, was a male. The order sought was granted.

In a more recent case[38] the parties were married in 1972; the marriage was consummated and the parties enjoyed normal married life. The plaintiff wife, 32 years old and an erstwhile male, sued her husband for divorce on the ground of his adultery with another woman. Adultery was proved but the validity of the marriage was contested. The plaintiff, in evidence, informed the court that she had been born and registered as a male and because of her transsexual tendencies, she had to undergo a sex change operation on psychiatric advice. According to her, the parties enjoyed normal marital life and the breakdown of the marriage had no connection with her change of sex. The court, however, relying on *Corbett*,[39] stressed biological factors in sex determination and declared the marriage a nullity, being one between two persons of the same sex.

This judgment from South Africa has been criticized by M.L. Lupton on the following grounds: First, psychologically, the plaintiff considered herself to be a female and since the physical effect of reassignment surgery alongwith re-registration as a female had also to be considered, the psychological factor should have been given due weight. Second, an over-technical interpretation leads to disadvantages and inhumanity from the vantage point of society in general, even if it be in conformity with the views of medical experts. Third, if the legislature allows re-registration, it should also specifically sanction the marriage of an individual who has undergone change and re-registered his sex.[40]

Mention may be made here of a Swiss case[41] where the court, confronted with a similar problem, held that psychological sex, coupled with surgical reassignment, should be accepted as the criterion. It held:

> Now that the patients' psychic association with the female sex is strongly supported by anatomical changes, it appears to us impossible to go back. It would therefore be advisable to recognise legally a state which the law did not prevent from coming into existence.[42]

In fact denying transsexuals the right to marry would be "condemning them to the twilight limbo which can only immensurably increase their unhappiness and frustration."[43]

In *B.* v. *B.*,[44] the Supreme Court of New York held that a marriage between a woman and a transsexual who was born a female but later had a sex change surgery before the marriage, was void. According to the court:

> Apparently, hormone treatments and surgery have not succeeded in supplying the necessary apparatus to enable defendant to function as a man for the purposes of procreation. In the same way surgery has not reached that point that can provide a man with something resembling a normal female sexual organ, transplanting ovaries or a womb. Those are still beyond reach.[45]

The marriage of a transsexual has been held to be valid, at least in part, by a US court in 1976.[46] It was a marriage between two biological males. The court said:

> [It was] impelled to the conclusion that for marital purposes if the anatomical or genital features of a genuine transsexual are made to confirm to the person's gender, psyche or psychological sex, then identity by sex must be governed by the congruence of these standards . . . If . . . sex reassignment surgery is successful . . . we perceive no legal barrier, cognizable social taboo, or reason grounded in public policy to prevent that person's identification at least for purposes of marriage to the sex finally indicated . . . Such recognition will promote the individuals quest for inner peace and personal happiness, while in no way disserving any societal interest, principle of public order or precept of morality.[47]

The marriage was thus held valid and the husband was ordered to pay spousal support to the wife.

Post-marital Sex Change

Equally complicated problems arise where a transsexual undergoes sex change surgery after marriage.[48] Suppose a husband changes his sex after marriage, what would be its effect on marriage? It is submitted that such a change should not, by itself, affect the continued validity of the marriage. But if, by reason of such a change, the parties cannot live happily together leading to breakdown of the marriage, divorce my be allowed. In an unreported English decision of 1958, a woman petitioned for divorce on the ground of cruelty. She had been married to her husband for eight years and also had a child by him. The husband had, however, resorted, under medical advice, to treatment for changing his sex, and the wife contended that the husband's conduct, as a result of which her health suffered, amounted to cruelty. The petition was dismissed on the ground that no cruelty in law was established.[49] It is, however, submitted that divorce should have been granted on the ground of breakdown of marriage in this case since the relationship between the parties could no longer have continued to be harmonious in the circumstances.

Effect of Sex Change on Custodial Rights

It is a well established principle that the prime consideration in determining custody cases is the welfare and best interests of the children. It would thus appear that transsexualism, per se, cannot be an evidence of unfitness of any parent. The issue came up for consideration in one American case.[50] A mother had been awarded custody after divorce. Some years later, she underwent a sex change surgery and remarried a woman. Thereupon the father filed an application for change of custody of his four teenaged daughters on the ground of sex change of the mother. Under the Colorado statute pertaining to custody, decree cannot be modified unless it is shown that there is a change in circumstance which would affect the interests of the child. The court held that there was "no evidence that the 'mother's' transsexuality adversely affected 'her' relationship with the children or impaired their emotional development."[51] The court accordingly ordered custody of the daughters to remain with the mother.

Effect of Sex Change on Employment and Educational Facilities

Sex change may create practical difficulties in regard to those spheres of employment or educational facility where there are restrictions on the employment or admission of person of a particular sex. For example in the female ward of a hospital, generally only female nurses are allowed. If a female nurse becomes a male, one cannot expect the authorities to continue her in service in the female ward. Same would be the case where, in an exclusively boys college, a boy undergoes sex change.

Change of sex also raises questions in regard to employment and educational facilities in the context of discrimination. There have been numerous cases in the United States where the petitions of transsexuals being adversely treated by reason of their sexual status have been turned down. An interesting case came up before the district court of California.[52] The plaintiff, a woman, was employed as a hemodialysis technician by the defendant, a medical centre. She informed the director of personnel of the defendant that she wished to undergo sex conversion surgery. Shortly thereafter, she was discharged from service on the ground that such a change might have a potentially adverse effect on the patients receiving treatment at the centre as also on the plaintiff's co-workers looking after those patients. Thereupon, the plaintiff sued her employer under the Civil Rights Act 1964 for injunction as well as damages on the ground that the employer had been guilty of an "unlawful employment practice" within the meaning of the Act. Under the Act, inter alia, it is an unlawful employment practice "to fail or refuse to hire, or to discharge, any individual or otherwise to discriminate against any individual because of such individual's sex."

The court, however, held that the Act did not embrace any transsexual discrimination or any permutation or combination thereof. The Act spoke of discrimination on the basis of sex and made no mention of change of sex or of sexual preference. Similar view was taken in *Powell* v. *Read's Inc.*[53] The facts, in brief, were as follows. Powell, a male and legally a male till the time of filing the petition, was engaged in a trial venture of living as a woman as a prerequisite to having a sex change operation. He applied for a job of a waitress at a store and got it but was dismissed on the very first day of the employment. According to him, the ground for dismissal was the supervisor's discovery through the report of a customer, that he was a man. It was held that there was no cause of action under the Civil Rights Act as this did not constitute discrimination.

A reference may also be made of another interesting case decided by the New York .Supreme Court.[54] Renee Richards underwent a sex change surgery to become a woman after being a ranked male tennis player in the men's age group of 30 and over. After the surgery "she" sought to enter the ladies' singles competition in the US Open. She was asked by the teems association to take and pass a chromosome test before such permission could be granted. The court held that this was "grossly unfair, discriminatory and inequitable and violative of their rights." It concluded that previously the defendants had relied on observation of primary and secondary sex characteristics to determine the sex of potential competitors; that the defendants had instituted the test,. which Richards would have failed, for the sole purpose of excluding her, and that she was now a female as her external organs, her appearance and her psychological status were that of a woman and her internal sexual structure was anatomically similar to that of a person born female who had suffered a total hysterectomy and ovariectomy.[55]

Conclusion

The above discussion reveals that there is a haze of uncertainty surrounding the legal status of transsexuals. There is no doubt that cases of sex change by surgical and hormonal treatment are bound to increase with the development of surgical techniques all over the world and with the people's awareness of such techniques.[56] As is evident from the case law and instances discussed above some countries are already faced with such problems and courts often find themselves in a dilemma in the absence of clear cut provision of law. Some countries have endeavoured to sort out certain issues connected with sex change by enactment of statutes. In India, though no legal case has come to notice so far but instances are not wanting where an individual has changed his or her biological sex.[57] Undoubtedly conversion to another sex must have created a number of problems for the transsexual but probably due to ignorance or inhibition, Indian courts have not so far been seized of any such issue. That does not, however, imply that our legislators could sleep over the matter till our judiciary is actually confronted with the issue. Serious thought should be given to the matter on the following lines.

Transsexuality should be considered and recognized as a medical problem requiring treatment like any other disease or ailment. If other methods of treatment fail and sex reassignment is the only practical remedy there should be nothing in law to prevent the doctor or the patient from going ahead with the surgery or other treatment. Proper safeguards, however, should be there before the operation is performed. Preoperation investigations should be made thoroughly and the patient should be apprised of the consequences of the initial problems he/she might have to face in the society, and so on. It is important that such operation should be permitted only on therapeutic grounds in case of genuine transsexuality and not on cosmetic grounds whatsoever. Needless to say that consent of the patient should be imperative. Once the operation is performed and the person enters into the new world of the other sex, it should be legally permissible to make consequential changes in documents relating to birth, nationality and identity, *e.g.*, birth records and certificates, passports and visas.

Consequences of sex change on marital life should be made clear. Where the change had been undergone before marriage, without concealment or misrepresentation of the material fact, the marriage should be regarded as valid. In other words, pre marriage conversion of one party with knowledge of the same to the other, should not be made a scapegoat for moving out of a marriage. When conversion is resorted to after the marriage and that leads to problems, divorce should be made available on ground of breakdown of the marriage.

Custodial rights of a parent undergoing a sex change should not be affected if there is nothing else to show that he or she is not a befitting custodian. It should not be forgotten that a person who has had a sex change has already gone through, and possibly is still going through, an emotionally traumatic experience in life. His assimilation into the new world should be facilitated by understanding, sympathy and societal acceptance. His looks and apparent sex should be the criteria for judging his sex for all practical purposes and there is no doubt that the legal recognition of the changed sex as the true sex of a person is the most important and essential factor in rehabilitating a transsexual emotionally, physically, socially and psychologically.

REFERENCES

1. The medical term for this is "gender dysphoria syndrome". See Melvin M. Belli, "Transsexual Surgery: A new Tort?", 17 *J. Family L.* 487 at 490 (1978-79); Jerold Taitz "The Legal Consequences of a Sex Change—A Judicial Dilemma", 97 *South African Law Journal (S.A.L.J.)* 65 (1980).
2. *The Transsexual Phenomenon* (1966).
3. Belli, *supra* note 1 at 489, quoting Morris from *Conundrum* 62 (1974). Morris was a trans-sexual and *Conundrum* is his autobiography.
4. David William Meyers, "Problems of Sex Determination and Alterations", 36 *Medico-Legal Journal* 174 (1968).
5. See Betty Liddick, "It's a Tough World for Transsexuals" *Sunday Standard* (23 April 1978) (New Delhi).
6. Taitz, *supra* note 1 at 65.
7. *E.g.*, the Roman Emperor Heliogabalus (Egagabalus) (204-220; the eighteenth century French statesman and diplomat Abbe' de Choisy; the famous French soldier, lawyer, ambassador and renowned fencer Chevalier d' Eon; and the British Governor of New York (1702-08) Lord Cornbury., See ibid.
8. Katherine S. Robbins, "Behind a Search for Identity: The Transsexuals' Quest", 85 *Sci. Digest* 74-79 (1979).
9. See Bartholomew, "Hermaphrodites and the Law", 2 *Malaya Law Rev.* 83 at 89 (1960), quoting Moore from *Embryonic Sex Hormones and Sexual Differentiation* 7(1947).
10. *The Evolution of Sex* 266 (1932).

11. See Bartholomew, *supra* note 9 at 94-95.
12. *Supra* note 9.
13. "Recent Developments Concerning the Criteria of Sex and Possible Legal Implications", 31 *Manitoba Bar News* 104 (1959). See also Douglas K. Smith, "Trans-sexualism: Sex Reassignment Surgery, and the Law", 56 *Cornell L. Rev.* 963 (1970-71).
14. Bartholomew, *supra* note 9 at 88.
15. Smith, *supra* note 13 at 967-68.
16. Id. at 970.
17. Ibid., quoting J. Randell, "Indications for Sex Reassignment Surgery" (unpublished paper presented at the First International Symposium on Gender Identity, London, July 1969; on file at the *Cornell L. Rev*).
18. Liddick, *supra* note 5.
19. According to the estimates made by the Erickson Educational (an informational clearing house on the subject) although many transsexuals staunchly insist that they have no regrets, only two out of ten make a happy adjustment. See *ibid.*
20. *The Times of India* (3 December 1979).
21. See Meyers, *supra* note 4 at 180.
22. See *B. v. D. Van Niekert*, "Sex Change Operation and the Law", *S.A. L.J.* 239 (1970), citing a decision of the *Oberlandesgericht* of Frankfurt, reported in (1969). Neue Juristische, Wochenschrift 1976–77.
23. See Meyers, *supra* note 4 at 178, citing *In re X*, 73 Sheriff Court Reports 203.
24. See Taitz, *supra* note 1 at 69.
25. See Smith, *supra* note 13 at 972; quoting from E. de Savitsch, *Homosexuality, Transvestism and Change of Sex* 104-05 (1958), citing *In re Leber* (1945).
26. See David Pannick, "Homosexuals, Transsexuals and the Sex Determination Act", *Pub. L.* 279 at 296–98 (1983), citing the case reported in Series A, No. 40, 3 E.H.R.R. 557 (1981).
27. Quoted in id. at 299.
28. See ibid.
29. See George J. Alexander, "Family Law" 18 *Syracuse L. Rev.* 383 at 394 (1966-1967), citing *Anonymous* v. *Weiner*, 270 N.Y.S. 2d 310.
30. Quoted in Smith, *supra* note 13 at 997–98.
31. See id. at 998.
32. Id. at 998-1000, citing *In re Anonymous*, 293 N.Y.S. 2d 834.
33. "Sex Offences: The medical and Legal Implications of Sex Variations", 25 *L & Contemp. Prob.* 292 at 307-08 (1960).
34. See Smith, *supra* note 13 at 995, citing *Ill. Ann. Stat.* ch. III. 5, and S. 71–17(I) (d) (Smith-Hurd Supp. 1970), *La. Rev. Stat. Ann.* 40: 336 (Supp. 1971).
35. [1970] 2 All E.R. 33.
36. Id. at 48–49.
37. See Taitz, *supra* note 1 at 73, citing *Jonker* v. *Jonker*, an unreported judgment of the WLD of 25 February 1970 reported in *The Star* (26 February 1970) and *Sunday Times* (1 March 1970) (Johannesburg).
38. See id. at 73, citing *W. v. W.*, 1976 (2) SA 308 (W).
39. *Supra* note 35.

40. "The Validity of Post-Operative Transsexual Marriage", 93 *S.A.L.J.* 385-90 (1976).
41. See Smith, *supra* note 13 at 971-72, citing *In re Leber* (1945).
42. Ibid.
43. Niekert, *supra* note 22 at 241.
44. See Pannick, *supra* note 26 at 298, citing the case reported in 355 N.Y.S. 2d 712 (1974).
45. *Ibid.* In *C and D*, an Australian court declared a marriage between a true hermaphrodite, who had undergone sex change surgery, and a female as void. See (1979) 53 *Aust. L.J.* 659.
46. See id. at 295-96, citing *M.T.* v. *J.T.*, 355 A 2d 204 (1976).
47. See id. at 296.
48. There are reports that some married transsexuals, after having undergone surgery, continued to live with their original wives as sisters. *Toronto Daily Star* (13 July 1968).
49. *Dolling* v. *Dolling* (1958) (unreported) Original petition reported in *The Times* (23 May 1958) and appeal proceedings reported in *The Daily Telegraph* (14 November 1958).
50. See Jeannine S. Haag and Tami L Sullinger, "Is He or Isn't She? Transsexualism: Legal Implications to Integrating a Product of Medical Definition and Technology", 21 *Washburn L.J.* 342 at 351 (1982), citing *Christian* v. *Randall*, 33 Colo, App. 129, 516 P. 2d 132 (1973).
51. Id. at 352.
52. *Voyles* v. *Ralph Davies Medical Centre*, 403 F. Supp. 456 (1975). See also, for similar views, *Grossman* v. *Bernards Township Board of Education*, 11 F.E.P. Cases 1196 (1976).
53. F. Supp. 369 (1977).
54. See Pannick, *supra* note 26, foot note 63 at 296, citing *Richards* v. *U.S. Tennis Association*, 400 N.Y.S. 2d 267 (1977).
55. Ibid.
56. According to a recent newspaper report, a dozen women, who once were men, recently gathered in Sydney for a public seminar on transsexuals. Transsexuals and other seminar participants made a demand for a planned "gender crisis centre" where homosexuals, lesbians, bisexuals, transvestites, transsexuals and their relatives could go for counselling. According to Ms. Roberta Perkins, an anthropologist born as a male, about 2,000 transsexuals lived in Australia, 200 out of them have had reassignment surgery. This was stated to be one of the highest numbers per capita in the would. It is estimated that at Melbourne hospital about 20 sex change operations were performed each year. Before the surgery, man is required to live for two years as a woman. See *The Hindustan Times* (11 February 1981). Another report in *Rand Daily Mail* (9 February 1970) states that during the last twenty years there have been over 1,000 known sex change operations in the world, 47 of them in Great Britain. See Niekerk, *supra* note 22 at 241. The number must have obviously multiplied immensely since then.
57. In 1971, one Munni Bai underwent sex change. She now calls himself as Munna Lal and is a police constable. In 1979, a 16 year old married girl, Sharda, from Sehore District of Madhya Pradesh, underwent an operation, became a male and now calls himself Sharda Prasad. See *The*

Hindustan Times (22 February 1979). In another instance, a 19 year old girl developed masculine features and was refused admission to a girl's college unless she produced a medical certificate to confirm her sex. See *supra* note 20.

24

HERMAPHRODITE MARRIAGE*

Legal reasoning and interpretation have made tremendous advancement during the last few decades. Nonetheless, it cannot be denied that it still lags far behind the techniques of medical science and there are many spheres where jurists, judges and lawyers still find themselves in a dilemma. An area which needs immediate reflection is in regard to the legal status of those human beings who live in a "twilight zone" being biologically neither purely male nor purely female. According to an eminent authority (Bracton) "Mankind may also be classified in another way; male, female or hermaphrodite . . . . A hermaphrodite is classified as male or female according to the predominance of the sexual organs".

In Plato's (the great philosopher) *The Symposium* it is suggested by Aristophanes that "originally there were three sexes, not as with us, two, male and female; the third partook of the nature of both the others and has vanished, though its name survives. The hermaphrodite was a distinct sex in form as well as in name, with the characteristics of both male and female, but now the name alone remains, and that solely as a term of abuse."

---

* Hindustan Times, February 23, 1984

There is still another category of human beings who belong, biologically, to a particular sex but psychologically to the other. In other words, the trait of a particular sex are predominantly prevalent in them but emotionally and psychologically they feel that their biological sex does not conform to their rightful gender. The problem is not as simple as it seems to be. Such persons are obsessed with a desire to change their sex by medical and surgical devices. They feel depressed, embarrassed and frustrated and often develop suicidal tendencies. They are known as transsexuals. Sex change by surgical operation is not a phenomenon in the air alone but quite an earthly problem. Though many countries have enacted special laws to govern their legal status on various matters, yet problems have arisen and courts have been unable to deliver justice in the real sense as distinct from justice according to the law.

Though our courts have not been so far confronted with problems pertaining to hermaphrodites and transsexuals (persons who have undergone a sex change), it is high time that thought be given to it before a situation actually comes up. According to a report, in India the population of eunuchs reportedly, is 12.5 lakhs. They are very much alive to the humiliation, discrimination and problems they live in and are now heading towards concerted moves to mitigate their sufferings. The other day, the Executive Committee members of the Delhi Pradesh Hijra (Eunuch) Kalyan Sabha sent an appeal to all world leaders, presidents, prime ministers, UNO, WHO and Human Rights Organisations to help them. Amongst their demands are free education upto higher secondary, vocational training, employment and old age benefits, etc. Apart from these demands, one is not too sure about the legal status of their marriage.

We have a good number of hermaphrodites and even transsexuals in our country. Our law and courts should be prepared to tackle litigations concerning them.

On the marital status of transsexuals there have been a number of decided cases abroad, like *Corbett* v. *Corbett* (English) 1970, *B.* v. B. (USA) 1974, *M.T.* v. *J.T*, (USA) 1976, and so on and verdicts have been mixed ones. Some courts stressed the biological factor in sex determination whereas others emphasised the psychological factor. The former interpretation has worked hardship as marriage between two persons of the biologically same sex has been declared void even though the transsexual belongs socially, mentally, psychologically and physically to the sex reassigned after medical intervention.

It would be of interest to note an Australian case on the point decided by the Family Court of Australia at Brisbane on April 20, 1979. The respondent (husband) was diagnosed as a hermaphrodite at the age of twenty-one. He married the petitioner, who, twelve years after the marriage, filed an application for a decree of nullity. The same was granted on two grounds, viz. the respondent's consent to the marriage was not a true consent because she was mistaken as to the identity of the other party at the time of the marriage — she believed that she was marrying a male whereas in fact she was marrying a combination of both male and female. The second ground was that the marriage is a voluntary union of a man and a woman and since the respondent was neither man nor woman but a combination of both, a marriage in the true sense could not have taken place and there was thus no marriage.

In this case biologically, the respondent's chromosome pattern was female but he possessed both male and female gonads. He had been brought up as a male. By surgical operation, his female genitals were removed and male genitals were fixed. According to the medical report itself "there is now nothing in the patient's appearance to distinguish him in any way from a normal adult male. He shows no personality disorders of any kind and is quite secure in his maleness. He is happily engaged and is shortly to marry the girl with whom he was in love before his surgery was commenced." Thus, it is very clear that the respondent was for all social and practical purposes a male and in love with the girl he married later.

It is also presumed that the petitioner was aware of the surgery and its purpose. It is unfortunate that under these circumstances the marriage should have been annulled *over a dozen years* after it was solemnised. Except for procreative function, the respondent was a male but the court decreed annulment saying that he was *neither a male nor a female.* It is submitted that such an interpretation and over technical view is cruel and hard for the not-so-lucky individuals. Every one has a right to marry and set up a family.

If the law does not assign any definite gender to a transsexual or a hermaphrodite, it would be denying him/her this basic right. Procreation is one of the most important objects and incidence of marriage, no wonder, but not the be-all or end-all of a marriage. After all, sterility is not made a ground for annulment or divorce. Where parties enter into a relationship fully alive to the facts and circumstances, an over technical interpretation of sex should not come in the way of normal life.

25

SEX OF THE ATHLETE*

A few decades back it was almost inconceivable that a person, apparently belonging to a particular sex, in fact medically belonged to the other; in other words, that a medically male could pass off as a female and vice versa without an iota of suspicion. Medical advancements have caused both clarity and confusion in this respect and given rise to many enigmatic situations.

Sex change surgery is no longer a utopian idea and many persons suffering a gender identity crisis have undergone this surgery. There are many cases of males becoming females and vice versa. Needless to say that their transformation from one gender to the other is not the end of their emotional and physical problems; it is beset with problems relating to rehabilitation and adjustment in the new role and their acceptance and recognition by the world in that newly acquired role. We come across many cases where transsexuals have fought their claims in courts in matters pertaining to matrimonial relations, employment, change of name and status in identity documents, and so on.

A matter that is allied and of interest is in regard to sex-determination in sports. Sports occupy a very important position not only at the individual, state, or national level but also

---

* *Hindustan Times*, September 6, 1984

at the international level. No wonder all arrangements are made and steps taken to make the events successful and pleasant without any element of deceit, foul play or fraud. Towards the achievement of this, amongst other things, is the sex determination test resorted to before such grand events. During the recent Olympics 2,500 women athletes competing in various games had to undergo a sex test to determine that they were, infact, women. The incidents that prompted the need for such curious test are extremely interesting, though embarrassing.

In 1964 during the Tokyo Olympics Ewa Klobukowaka of Poland competed in the 100 metre dash and won the bronze medal. In 1965 she failed a chromosome test before a championship meet and was barred from international competition. She was also stripped of all her records. Ever since this incident, women athletes have had to pass through a gender clinic for sex verification before participating in the games.

Some players find it embarrassing and have chosen not to play rather than submit to this scrutiny. Tamara Press of Soviet Union, who had won a gold medal in 1960 Rome Olympics and the Tokyo Olympics in 1964, and her sister Irina who came first in the 80 metre hurdles in Rome, have refused to submit to the test and hence to play also.

A reference may be made to an Indian girl, an athlete aged about nineteen, who was refused permission to participate in a kabaddi tournament, a game in which she had won laurels in the past, because she had started showing signs of transformation by developing masculine features like moustaches, etc.

In one case from America, Rennee Richards, a ranked male tennis player in men's age group of 30 and over, underwent a sex change surgery to become a woman. After the surgery 'she' sought to enter the ladies singles competition in the US Open. The US Tennis Association asked her to take and pass a chromosome test before such permission could be granted. The court found that this was "grossly unfair, discriminatory and violative of her rights". It concluded that previously the Association had relied on observation of primary and secondary sex characteristics to determine the sex of potential competitors; that the defendants had instituted the test — which Richards would have failed — for the sole purpose of excluding her; and that she is now a female "as her external organs, her appearance and her psychological and social status — were that of a woman and her internal sexual structure was anatomically similar to that of a person born female who had suffered a total hysterectomy and ovariectomy" (*Richards* v. *US Tennis Association*, New York Supreme Court, 1977).

However, the most interesting episode is that of Hilda Strike. Hilda Strike, a 74-year-old Canadian grand-mother now, had participated in the 1932 Olympics. She narrowly lost by inches, the 100 metres to Stella Walsh a Polish born American. Walsh died in 1980 and the autopsy revealed that she was a man and not a woman. Hilda now wants her gold medal. John Holt of the International Amateur Athletic Federation observed "on the basis of this reported inquest on the late Stella Walsh, Hilda is quite entitled to consider herself to have been the world's fastest woman at the time." However, no action is possible now in this case since the event happened long back and the discovery was made only after the death of the gold medalist; nonetheless, certain issues do crop up in such cases. Would Hilda have been entitled to have her gold medal back had Walsh been alive and the discovery made during the lifetime? Does it amount to fraud; if so with what consequences? What would be the position in case of trans-sexuals such as the one mentioned above (Richards' case). Would the authorities insist on chromosome tests or simply on observation of primary and secondary sex characteristics in such cases? If so, can the authorities discriminate between trans-sexuals and other players in the nature of the test? Also, what would be the position in case of persons (like the case of the Indian girl mentioned above) who are undergoing a process of transformation without surgery. And last but not the least, one could ask, why the scrutiny only in case of female players. May be medical science has a convincing answer to this — a man's identity can never be in doubt; he is blessed with precision and clarity; only females are capable of being vague and uncertain even when it comes to such a thing as gender identity. What a discrimination by nature also, if at all!

26

M.P. COPS 'MARRIAGE'*

The "marriage" of two women constables of the Madhya Pradesh Special Armed Force (SAF) with each other and their subsequent discharge from service last month has created a stir in society. The two women entered into "wed-lock" with the consent of their parents. They are dismissed from service because their "marriage" became an embarrassment for the police and the action is being defended by some officials on the ground that the "wedding" amounted to indiscipline.

Both the 'spouses' had earlier married hetero-sexually. The 'male' counterpart is separated from her husband and is said to have had 'relationship' with other women in the battalion earlier. The 'wife' is a widow with three children. Their families do not disapprove of their relationship. Infact, they dressed them up as bride and groom and had them photographed. In one of the photographs, the 'wife' is shown touching the feet of the 'husband'.

The couple are puzzled over the hue and cry. According to them "you have no idea what life is like inside our barracks. There are 90 women in our company and usually a woman will be more friendly with someone. A good many of them have paired off in our company." Marriage, according to them, "is a wedding of two souls — where in the scriptures is

* *Hindustan Times*, March 26, 1988

it said that it has to be between a man and woman"? In fact, some officers in the police force claim that the "marriage" was a kind of protest against society which treated them badly when they were single and lonely.

Whatever it is, the 'marriage' has raised a sociological and legal problem. The pertinent issues are: Can there be a 'marriage' between two persons of the same sex? What would be the legal implications if the two undergo a ceremony or ritual of a form of marriage? Would such an act amount to an act of moral turpitude or indiscipline so as to invite their dismissal or disciplinary action against them? If so, how do we reconcile this with a person's right to liberty, for, after all, cannot two consenting adults have a right to live in an emotional union without causing any detriment to anybody else? These and a host of other issues revolve around such 'situations'.

Living together of two persons of the same sex need not necessarily be a sexual bond. The link could be more emotional. However, it is often assumed that the roles the two partners play in this intimacy includes participation in lesbian or homosexual acts. Researches in penal and other institutions reveal that "homosexuality in female institutions tends towards less predatory forms". Some girls admit to be "going with" or "married to" another girl. Amongst imprisoned women especially, the tendency to group themselves into "make believe" families which may include their playing the role of a 'husband' or 'wife' has been noted. Loss or lack of support from their families or groups outside, which is believed to be particularly hurtful to women, has been held in part to be responsible for these quasi-kinship groups. This seems to be the situation in the present case as well.

The question is, even if one were to concede that there is something beyond a mere emotional tie, can we punish such behaviour? In fact homosexuality involves not so much a choice to act wickedly as the seeking of fulfilment in an abnormal way. This is a symptom of some emotional disturbance and problem in personality development, which needs a sociological and psychological rather than penal handling. Coming to the status of the 'marriage' in question since such a union is not capable of being consumated, the marriage would be void under the marriage laws. Medical advancement has made sex change by surgery or therapeutic devices possible. This has sometimes led to problems in marital relationship. In the noted English case, *Corbett* v. *Corbett* (1970, England), the respondent, was a male at birth. He underwent a sex change operation which meant removal of male genital organs and providing of artificial female organs by plastic surgery. He married the petitioner who was a male. After detailed consideration of medical evidence, the court came to the conclusion that biological sex of a person is fixed at birth and cannot be subsequently changed by artificial means. In view of this, the court held that the marriage was void. Thereafter, the Matrimonial Causes Act, 1973 lays down specifically that, of the two parties to marriage, one must be a male and the other female. Thus, it would seem that there can be no marriage between two males or two females.

It is thus clear that a marriage in order to be a legal bond must be a hetero and not a homo sexual union. Thus the marriage of the two women constables is not a valid relationship in the eyes of the law. It is at best a companionship, an emotional bond which the parties have entered into because of their liking, weakness, attraction or sympathy for each other. Coming to the question of 'morals', 'morality or ethics' of the situation one is reminded of the following quotation: "Legislators should stop moralising through the sanction of criminal law and remember that even Jesus Christ gave man the free will to commit immoral acts. Criminal statutes ought to be strictly tailored only to particular behaviors that offend the lives and property of others". (Orloski, *Criminology*).

On the touchstone of the Penal Code too, one would find it difficult to treat it as an offence. Nor does the contention that the women have committed an act of indiscipline by the so-called marriage sound to reason.

SECTION VII

DEATH AND DYING

27

IS ARTIFICIAL LIFE BETTER THAN NATURAL DEATH?*

Karen Quinlan, a 22-year old American girl, lapsed into coma on April 15, 1975, with irreversible brain damage. An artificial respirator was used to sustain her breathing. She was kept alive in a hopless, vegetative state for a couple of months on that respirator. Her father successfully moved the court to enable him to authorise the removal of the respirator. The court held: "No external compelling interest of the state could compel Karen to endure the unendurable, only to vegetate a few measurable months with no realistic possibility of returning to any semblance of cognitive or sapient life."

Such cases are bound to come up with the development of sophisticated life prolonging devices which in fact can do no more than artificially maintain breathing after most other bodily functions have ceased. Under these circumstances, certain questions arise. Should the patient be compelled to linger on? How long should he or she be kept artificially alive? Can the patient ask for removal of these devices so that nature can have its course?

Suicide apart, self-death has been practised since long. It is believed that African bushmen and some primitive tribes in North America left their infirm elders in the wilderness to die.

---

* *Indian Express*, September 5, 1979

Sometimes they were given poison. There are references to assisted self-death to mitigate pain in St. Thomas More's *Utopia.* Francis Bacon in his *New Atlantis* says it is the physician's responsibility not only to aid recovery but also at times to "make a fair and easy passage".

Among recent cases, in May, 1962, Mrs. Suzanne Vande gave birth to a "thalidomide" baby in Leige, Belgium. Eight days later, she mixed a lethal dose of sedative in the infant's feed, and the baby died. She, along with her husband, mother, sister, and the family doctor were tried but acquitted after the jury was told of a public referendum which scored 16,732 to 938 votes in favour of acquittal.

In 1950, Carol Paight, a 21-year old college student, shot her father in his hospital room shortly after it was discovered that he had terminal cancer. She was acquitted on ground of temporary insanity. Braunsdorf shot his 29-year old spastic daughter as he was worried about her future after his death, and himself attempted suicide, but failed. He too was acquitted on ground of temporary insanity.

The few cases that have been reported in the other countries involved active mercy killing like administering an injection, shooting, etc. But the sympathy of the doctors and the public is for permitting nature to take its course, thereby shortening the dying process; this is distinct from active killing.

Those against withdrawal of life support devices hold that any intervention that shortens life is morally reprehensible. Besides, they say even terminal diseases can be cured under certain circumstances. They also feel that a patient's consent to mercy killing may not be voluntary in the real sense of the term. Such a person is in great physical and emotional agony and perhaps under the influence of drugs. If the authority is given to others, there is a likelihood of abuse.

Others, however, argue that the patient is relieved of the indignity of living devoid of the exercise of his faculties or self-control. "To prolong life uselessly while the personal qualities of freedom, knowledge, self-possession, and control and responsibility are sacrificed, is to attack the moral status of a person." (Fletcher, *Morals and Medicine*, 1954).

Another doctor says, "Where there appears to be no possible chance of return to any type of conscious awareness, much less any comfortable existence, the act of omitting tube feedings is good medicine." (Williamson, "*Prolongation of Life, or Prolonging the Act of Dying?*", 1967).

By seeking the Supreme Court's permission to switch off the respirator which had held his daughter's breathing process, Mr. Quinlan acted most wisely. Apart from being a financial strain, prolonging the existence of a person who can never have a human existence is agony to both himself and others. Besides, when it has been established beyond doubt that the patient's condition is irreversible, it is better to use hospital resources for treatment of hopeful cases.

In India there is no law permitting mercy death and there is not a single case on the point. It would appear, however, that withdrawal of life saving device or administration of something leading to death, if done with the consent of the deceased if he is above 18, is not murder. Nevertheless, it will be culpable homicide.

Culpable homicide is defined thus: "Whoever causes death by doing an act with the intention of causing death, or with the intention of causing such bodily injury as is likely to cause death, or with the knowledge that he is likely by such act to cause death, commits the offence of culpable homicide."

There are five exceptions under which culpable homicide will not amount to murder. The fifth says: "Culpable homicide is not murder when the person whose death is caused, being above the age of 18 years, suffers death or takes risk of death with his own consent."

There was a proposal in 1973 to provide for mercy killing of persons who have become completely invalid or those suffering from an incurable disease. A Bill to that effect was introduced in the Lok Sabha on December 14, 1973 but was shelved.

Whereas killing by positive acts, even as an act of mercy, may shock the conscience of many, yet, unnecessary prolongation of a life which has been afflicted by deadly disease or accident as a result of which the patient can have no more than a mere mechanical existence, should not be encouraged especially if the patient consents to withdrawal of such treatment.

There should be a limit on how far mechanical devices should be used for the sustenance of a mere breathing process, which is no blessing for the patient and no solace for the relations. It's a source of distress and a misuse of medicine.

It is, therefore, desirable that with proper scrutiny and safeguards, a person should be permitted to cut short the agony of a prolonged dying process and the persons assisting should not be held criminally liable.

28

SUICIDE VS. LAW*

While there are serious debates on legalising euthanesia on one hand, attempt to commit suicide continues to be an offence under our penal laws. Section 309 of the Indian Penal Code says, "whoever attempts to commit suicide and does any act towards the commission of such offence, shall be punished with simple imprisonment for a term which may extend to one year or with fine or with both." Thus attempt to commit suicide is an offence entailing prosecution and punishment. Prior to 1882, it was compulsory upon the court to impose a sentence of imprisonment — however short — in such cases. After an amendment the court has been vested with a discretion and now the condemned survivor may just get away with some fine.

Suicide, an act of voluntary and intentional self-destruction, has been viewed with horror and revulsion by some societies whereas others have extolled it. The *Koran* condemned it; Plato felt that, "They must have no compassions whatsoever in the tomb, for they must be buried ignominiously in waste and nameless spots . . . and the tomb be marked with neither headstone nor name." (*The Laws IX*. 073). Plato, however, made an exception in case of those who took their own lives through intolerable stress.

---

* Hindustan Times, January 12, 1986

Suicide was also condemned by Aristotle, and under the Roman Law. According to St. Thomas Acquinas, God alone had control over life and death, and one who decides the moment of his own death and commits or attempts suicide, usurps God's power.

In India, the Brahmins honoured the person who "voluntarily freed his body." *Sati*, a widow immolating herself on the husband's pyre, was considered to be an act of great virtue. The Greek permitted criminals who were convicted of certain offences to put an end to their lives.

Even today many countries do not consider attempt to suicide as a crime. Under the English law it was an offence to attempt suicide, punishable by fine or imprisonment. The law was changed in 1961 and attempted suicide is no longer an offence.

In the United States, under the laws of most of the States, it is not a crime.

Under the Indian Law, as already mentioned, the Indian Penal Code, vide section 309, makes attempt to suicide an act of crime. How far is it just to punish a person who survives an unsuccessful attempt to finish his life? Those who advocate retention of the crime of attempted suicide argue that punishment would act as a deterrent. Such argument, however, is illogical. Would not a person make fool-proof arrangements to see that he does not survive? And if unfortunately he does wouldn't prosecution and conviction add to his miseries? Nothing can prevent him from making another attempt. If, on the other hand the law is understanding, and sympathetic it can bring hope in the lives of such persons.

Laws all over the world are known for their technical interpretation. This mode of administration of justice has worked great injustice in many cases. Justice Sachar former Chief Justice of the Delhi High Court, deviated from the strict interpretation of law in early 1984 (*State* v. *Sanjay Kumar Bhatia* March, 1984). In this case, a young man made an unsuccessful attempt to take his life by consuming Tik-20, because of extreme, frustration in an affair. Justice Sachar made a strong plea for the abolition of the crime of attempt to commit suicide. He observed, "It is a strange paradox that in the age of votaries of euthanesia, suicide should be criminally punishable. Instead of the society hanging its head in shame that there should be such social strains that a youngman should be driven to suicide, it compounds its inadequacy by treating the boy as a criminal. Instead of sending the young boy to a psychiatric clinic, it gleefully sends him to mingle with criminals, as if trying its best to see that in future he does fall foul of the punitive sections of the penal code."

Then on December 13, 119 cases of attempted suicide pending in the criminal courts were collectively quashed by the judge. The Delhi Administration and the police have been directed not to prosecute such victims in future. This was done under section 482 of the Criminal Procedure Code which empowers the High Court to quash a case if it feels that the continuance of a prosecution would not be in the interest of justice.

A person who restorts to the extreme step of taking his own life should be viewed as a patient in need of psychological and emotional care rather than as a criminal deserving punishment. In our society many people are driven to suicide because of economic insecurity, marital disharmony, emotional imbalance, etc. Local hospitals, psychiatric and health clinics and social welfare agencies can provide effective counselling. Associations like the American "Save A Life League", the English 'The Samaritans' and the Swedish emergency clinics for depression should be set up. In fact people with suicidal symptoms and tendencies should be treated with love and understanding and not punished as a matter-of-course.

29

LIVING WILL STATUTES*

Euthanasia or mercy killing by withdrawing of artificial respirators or withholding treatment, or administering something at the instance of friends or family members or doctors is bound to raise ethical and legal issues, and opinion on the desirability of a legislation permitting it is sharply divided.

A solution for those who wish to avoid the agony of prolonged suffering by use of life support procedures is the drafting of a living will. Under such a will, a competent individual may execute a declaration saying that if he becomes terminally ill or goes into an irreversible coma, artificial life sustaining treatments that only prolong the dying process, should not be used. There are many states in America which have passed statutes enabling individuals to execute living wills.

In the context of living wills, a very pertinent issue arises is, what if the woman executing a living will is pregnant at the time the will is to take effect? In most of the states of America, the living will statutes contain a pregnancy clause specifying that the terminally ill or comatose patients' living will is suspended if the patient is pregnant. The majority of these statutes suspend the will throughout the pregnancy while a few suspend it only if the patient is carrying a foetus that

---

* *Hindustan Times*, October 28, 1991

could develop to the point of birth if the mother is kept on life support procedures. Now that abortion is legalised in many countries, including India under certain conditions, how would these pregnancy clauses fare on the touchstone of legality?

This issue has to be analysed in the light of the famous American case, *Roe* v. *Wades* (1973). In this case the Supreme Court struck down a Texas statute criminalising abortion. A woman has a right to privacy inasmuch as she is free to choose whether to bear a child or terminate the pregnancy. This right, however, like any other fundamental right is not absolute. The state may intervene when it has compelling interest to protect and the two state interests thus identified were — maternal health and the potential life of the foetus.

There is very little case law directly on the subject of pregnancy clauses in living wills because, first, living wills are relatively recent and besides, there are few instances in which a terminally ill or comatose pregnant woman has executed a living will. There arc some cases, however, in which courts in America have tangentially discussed the issue. In *University Health Services* v. *Piazzi* (1976) a hospital's petition for continuing life support systems for a brain dead pregnant woman was granted even though her husband and family objected to it. The lady did not have a living will nor did the court attempt to find out what her wishes would have been had she been in health. It was argued for her that her right to privacy — which included both the right to refuse treatment and the right to terminate pregnancy— mandated removal of life support systems. The court did not accept the argument and held that as she was dead as defined by Georgia law, she had no protectable privacy interest. Besides, even if she had a living will, that would have been declared as suspended under the Georgia living will pregnancy clause.

In a recent case, *in re A.C.* Columbia (1987) a competent but terminally ill woman was 26 weeks pregnant when a tumour was diagnosed in her lung. She was given just a few weeks to live. After discussion with her doctors and family, she agreed to undergo a treatment that might extend her life until the time foetus was 28 weeks old, (chances of viability being greater at that point). When this plan was communicated to the hospital, a court order for a caesarean was sought. Attorneys to represent the lady and the foetus were appointed and a hearing held at the hospital. The court found that the foetus was viable and that "there was a state interest in protecting the potential life of the foetus." Initially, the lady agreed but after a few minutes refused. When the surgery was scheduled, appeal was made on telephone for a stay. The request was turned down and the surgeons performed the operation the same evening. The foetus was not viable and survived for only two hours. The lady died two days later. The court, while recognising that "we all know that we may have shortened A.C's life span by a few hours," sought to justify its order by discussing "the importance of balancing between the individual's right to bodily integrity against the state's interest in potential human life."

In this backdrop, it is submitted that in order to put an end to the suffering of those languishing in a comatose or vegetative state or surviving only on life saving devices, living will statutes should be enacted. An individual should have a right to die in dignity in just the same way as he has a right to live in dignity.

As regards pregnant woman, a pregnancy clause which suspends the operation of a living will without regard to the stage of pregnancy is unreasonable. It is submitted that a woman who is medically in a bad condition should have the same right to have her pregnancy terminated as a woman in good health.

30

SUICIDE: SANS DETERRENCE*

Today in many countries attempt to commit suicide is not a crime; abatement or inducement to commit suicide, however, is. Under English law prior to 1961, attempted suicide was a misdemeanour punishable with fine or imprisonment. There were strong opinions against it.

Under the Suicide Act, 1961, attempt to commit suicide is no longer a crime under the English law. In the United States and USSR attempted suicide is not a crime. Even in Japan where the custom of compulsory Seppuku (hara-kiri) or ritual of embowelling as a mode of punishment was made illegal in 1868, voluntary hara-kiri as an act of self-defeat or self-sacrifice has always been highly praised. In India, Section 309 of the Indian Penal Code could penalise an attempt to commit suicide with imprisonment for a term which may extend to one year or with fine, or with both. It is pertinent to note that under the original sections, prior to 1882, a sentence of imprisonment, however short, was compulsory. Now the court has discretion and may just impose a fine.

The serious debate on whether to decriminalise suicide or retain it in the penal code has at last been put an end to at least legally, and perhaps *only* legally, by the Supreme

---

* *Hindustan Times,* May 9, 1994

Court judgment *in P. Rathinam/Nagabushan Patnaik v. Union of India* (April 26, 1994). The history of this case is as follows:

As early as 1971 the Law Commission took up this issue and recommended that the penal provision of Section 309 is harsh and unjustified and so should be repealed. In 1979, a Bill was introduced in the Rajya Sabha seeking to delete Section 309 but it lapsed on the dissolution of the Lok Sabha soon after. In 1984, Justice Sachar (the then Chief Justice of the Delhi High Court) collectively quashed 119 cases of attempted suicide pending in the criminal courts and directed the administration and the police not to prosecute such victims in future.

In 1985, the same judge upheld the acquittal of an accused charged under Section 309 (*State* v. *Sanjay Kumar Bhatia*). He observed: "It is ironic that Section 309 IPC still continues to be on our penal code. The result is that a young boy driven to such frustration so as to escape one's own life would have escaped human punishment if he had succeeded but is to be hounded by the police, because attempt was foiled. Strange paradox that in the age of votaries of euthanasia, suicide should be criminally punishable . . . The continuance of Section 309 IPC is an anachronism unworthy of a human society like ours. Medical clinics for such social misfits certainly, but police and prison, never."

In *Maruti Shripati* v. *State of Maharashtra* (1987, Bombay), the provision was challenged as being violative of Article 21 of the Constitution. A constable was facing prosecution for attempting to immolate himself outside the Commissioner's office. The court accepted the petition filed by his wife challenging the constitutional validity of Section 309. It observed: "There is nothing unnatural about the desire to die and hence with the right to die. Whatever the circumstances which induce a person to end or terminate his or her life, the act of termination of life is the act of that individual. It is no less his than his act of living."

It was held that under Article 21, the right to life includes the right not to live or not to be forced to live. So far as the deterrent value of punishment in these cases is concerned, the court held that in case of persons with mental disorders, what is needed is psychiatric treatment and not confinement in prison cells which is going to make their condition worse. Those attempting suicide on account of acute physical ailment etc. need nursing homes and not prisons to prevent them from making the attempt again. In case of those who want to die for social or political cause or because of loss of interest in life, no deterrence can hold them back from attempting again.

The Andhra Pradesh High Court, however, took a different stand in a case before it, in 1987 (*C. Jagadeeswar*). The appellant, who was a doctor, had killed his four minor children by administering some poisonous tablets and later attempted to commit suicide alongwith his wife. The sessions court convicted him for murder and for attempted suicide. In appeal, the Andhra Pradesh High Court observed: "Unless a man is assured of physical existence there can be no other fundamental right." And further, "if attempt to commit suicide is not an offence, people who actively assist and induce persons to commit suicide may go scot-free."

The constitutional validity of Section 309 was accordingly upheld. Writ petitions were filed before the Supreme Court on the issue of validity or otherwise of Section 309. The prosecution for the petitioner who was arrested for going on a hunger strike and was being tried in a court in Orissa was stayed.

The Supreme Court, by its judgment, has set all doubts and uncertainties at rest by holding against the constitutional validity of Section 309. In a well-researched judgment, the court has discussed the issue on the touchstone of constitutionality, morality, religion, ethics, public policy, secularism, expediency, sociology and humanitarianism. The position in other countries has also been discussed. Holding Section 309 as violative of Article 21, it has held that "Section 309 of the penal code deserves to be effaced from the statute book to humanise our penal laws."

This judgment may have come as hope of liberation to some who are in a very depressed and irreversible state of mind or body, without fear of any punishment should their attempt fail, but what about its effect on young, immature minds who tend to act or react in desperate haste? For such people probably the strongest deterrent against the extreme step would have been the penal provision. Some frustrations and depressions in life could be just passing phases like failure in examination, in love, in getting a job etc. and a person who has no one to give him solace at that particular time might take the extreme step without fear of any law.

Though the fear of a penal law is not the ideal force, yet so long as we do not have adequate services by way of organisations or societies to give solace, counselling and requisite help to the victims of frustration, we need some authority at least to check unbridled attempts to commit suicide. In any case all situations cannot be treated alike. Neither should all victims of such attempts be absolved, nor all such victims deserve to be penally treated. What is needed is humanising the judicial approach rather than humanising a law by deleting a section without regard to subjectivity.

31

SUICIDE: A JUDICIAL RETREAT?*

"You ever gentle gods
Take my breath from me!
Let not my worser spirit
Tempt me again
To die before you please"

— *Shakespeare*

There has been a prolonged debate on the issue whether an attempt to commit suicide should be retained in the penal code or taken out of it. Most of the countries have depenalised such attempts. In India, however, an attempt to commit suicide is (was?) an offence under Section 309 of the Penal Code which carries a sentence of imprisonment for a term which may extend to one year along with fine in some cases.

Over two decades back, the Law Commission recommended the repeal of Section 309. It was followed by a Bill to that effect which lapsed. Thereafter the issue became a matter of judicial adjudication in various high courts, with divergent opinions. Hence there was no settled law on the point until the Supreme Court judgment in the *P. Rathinam Nagabhushan Patnaik* case (April 26, 1994). The Court, in a well researched judgment, discussed the issue on the touchstone of constitutionality, morality, religion, ethics, public policy, secularism,

* *Hindustan Times*, September 27, 1994

expediency and humanitarianism. A reference was also made to the law on the subject in other countries.

Without going into the history and entire debate and arguments of the court on the issue, it is pertinent to note the Supreme Court virtually bull-dozed Section 309 of the Penal Code which penalises attempted suicide. It held "Section 309 of the Penal Code deserves to be effaced from the statute book to humanise our penal laws. It is cruel and irrational provision, and it may result in punishing a person again (doubly) who has suffered agony and would be undergoing ignominy because of his failure to commit suicide . . . . May it be said that the view taken by us would advance not only the cause of humanitarianism, which is a need of the day but of globalisation also, as by effacing Section 309, we would be attuning this part of our criminal law to the global wave length."

However, just four months after this judgement declaring Section 309 of the Penal Code as ultra vires, the apex court has upheld a conviction for attempted suicide (*Vijay* and *Gianchand Jain* v. *State of Madhya Pradesh,* September 2, 1994). The facts of this case, in brief, were as follows. The accused was convicted and sentenced by the Sessions Judge, Shahjapur, for murdering his wife and later attempting to commit suicide. He filed an appeal to the High Court (Jabalpur Bench) contending that he was not even present in the room when the murderous attack was made on his wife, but confessed that when he saw her in that state he was too upset and tried to end his life by hanging. These arguments were not accepted and the court found that the appellant, who was in a deep financial crisis had decided to kill his wife and thereafter end his own life. This was also supported by a suicide note which was found immediately after the occurrence at the place of murder. The High Court, accordingly, dismissed the appeal and upheld the conviction and sentence of the accused, who was sentenced to imprisonment for life for killing his wife and to rigorous imprisonment for one year for attempting suicide. Both the sentences were to run concurrently.

The appellant appealed against this to the apex court which was dismissed. The court discussed the issue of tenability of conviction for murder without even specifically touching Section 309, viz conviction for attempted suicide. It simply held, "we find no reason to interfere with the order of conviction and sentence passed against the appellant" which obviously includes upholding conviction for attempted suicide. It is significant to note that though the appeal was filed when the judgment on suicide had not come and the law was indecisive, but it was disposed of after the Supreme Court judgment. Thus the Supreme Court overturned its own verdict on the criminality of attempted suicide, which it had very clearly and firmly abolished a few months earlier. Though there would not have been any real or practical affect on the sentence of the accused in this particular case if the Court had sifted the "offence" of attempted suicide and acquitted him for that act, since he was in any case sentenced to life imprisonment, except that he is to serve one year of rigorous imprisonment for attempted suicide, yet, the court should not have overlooked this aspect; more so, since the issue came up before the highest court for the first time after the judgment.

What would be the impact of this judgment on the issue of attempted suicide? Both the cases (*P. Rathinam*, April 1994 and *Vijay*, September 1994) have been decided by a bench of two judges. Can a bench of two judges review or overrule a judgment of the same court, also delivered by a bench of two?

Under Article 141 of the Constitution "The law declared by the Supreme Court shall be binding on all courts within the territory of India." The idea is to have certainty and clarity. Though this does not bind the Supreme Court and "all courts" do not include the Supreme Court, nonetheless, the apex court can depart from a previous decision of the court only under certain situations, e.g. where it is satisfied of its error and its baneful effect on the general interest of the public or when there are special cirucumstances which call for a different interpretation, and so on.

If a bench is unable to agree to an earlier decision of another bench of equal number, the matter should be referred to a larger bench. In order to avoid vagueness and confusion it is essential that a clarification be made on the point expeditiously either by a larger or full bench suo motu, or by legislative amendment.

32

FOR A HUMANE APPROACH TO SUICIDE*

Our time is fixed, and all our days are numbered,/How long, how short, we know not — this we know,/Duty requires we calmly wait the summons/Not dare to stir till heaven shall give permission", writes Robert Blair in *The Grave.*

This is the tenor of the Supreme Court verdict in *Smt. Gian Kaur* v. *State of Punjab* on suicide, delivered on March 21, 1996. When the two bench judgment in *P. Rathinam Nagabhushan Patnaik* v. *Union of India* came, last year, it was felt that the debate on whether to decriminalise suicide or retain it on the Penal Code had been, at last legally, put to an end. But that was not to be. The same has now been set aside by a five-judge Constitution Bench of the Supreme Court which has held that section 306 (which penalises abetment of suicide) and section 309 (which penalises attempt to commit suicide) are not violative of Article 21 of the Constitution (right to life).

It may be recalled that there was a long history behind the *P. Rathinam* verdict. The Law Commission, as early as 1971 recommended that the penal provision of Section 309 was harsh and unjustified and so needs to be repealed. A Bill introduced in the Rajya Sabha in 1972 seeking deletion of the

---

* *Indian Express*, April 2, 1996

section lapsed. In 1984 Justice Sachar (the then Chief Justice of Delhi High Court) quashed 119 cases of attempted suicide pending in the criminal courts and directed the administration and police not to prosecute such victims in future. In 1985 the same judge upheld the acquittal of an accused charged under Section 309 (*State* v. *Sanjay Kumar Bhatia*). He observed "The continuance of Section 309 IPC is an anachronism unworthy of human society like ours. Medical clinics for such social misfits certainly, but police and prison, never."

Writ petitions were filed before the Supreme Court on the issue of validity or otherwise of Section 309. And then came the culmination point. The Supreme Court held against the constitutional validity of Section 309. In its judgment, the court discussed the issue of suicide on the touchstone of, inter alia, constitutionality, morality, religion, ethics, public policy, expediency and humanitarianism. It also referred to the existing position in other countries. After a detailed discussion and analysis, the court came to the conclusion that "Section 309 of the Penal Code deserves to be effaced from the statute book to humane our penal laws. It is a cruel and irrational provision, and it may result in punishing a person again (doubly) who has suffered agony and would be undergoing ignominy because of his failure to commit suicide."

This judgment was hailed by many as remarkable and humane but was assailed equally, as well. There is no doubt that de-penalisation of suicide has the potential of creating several tricky situations apart from the technical issue of retaining abetment and assistance of an act as an offence when the main act is no longer an offence under law. Besides many dowry death cases which could be instigation or abetment to commit suicide by girls could be manoeuvred or interpreted as simple suicide by the girl.

The law needs to be humanised no doubt but it would be dangerous to give a licence to everyone to end his/her life irrespective of the situations and circumstances. There should not be a blanket protective shield behind every attempted suicide. What is needed is a humanising of the judicial approach rather than humanising a law by deleting a section without regard to subjectivity. The courts must not ignore the circumstances, the stress and strains, that compelled a person to take the extreme step; but nor should it convey an impression that, "it is my life and I can end it whenever I wish."

The present judgment which arose out of a set of appeals challenging the constitutional validity of Section 306 (abetment to suicide) in view of the striking down of Section 309 in *P. Rathinam* held that Article 21 is a provision guaranteeing protection of life and personal liberty and by no stretch of imagination can extinction of life be read or included in protection of life.

A person has no right to take his own life. It is, however, important that the state should not ignore its legal, constitutional and moral duty to enable and assist people to live a life of dignity.

SECTION VIII

TRADING IN ORGANS/CORPSE

33

DEAD SERVE THE LIVING*

Modern science has made possible the enucleation of certain organs from the body of the dead and their grafting or transplantation in the living. This not only provides a fresh lease of life to the living but also a meaningful death to the deceased.

There have been cases of heart transplantation but these are few and far between. What is becoming more popular is the use of eyes and ears after death. In our country also people have started thinking on the lines of donating their organs after death and there are laws regulating this, so as to avoid legal complications for the medical experts and also to ensure that the task is done by experts especially skilled for the job. There are two very recent statutes, the Eyes (Authority for use for Therapeutic Purposes) Act, 1982, and the Ear Drums and Ear Bones (Authority for use for Therapeutic Purposes) Act, 1982, which provide the terms and conditions under which a person may donate his eyes and ears.

In 1957 the State of Bombay made a law known as The Bombay Corneal Grafting Act, 1957, which made a provision with respect to the use of the eyes of deceased persons for therapeutic purposes. This Act was extended to the Union Territory of Delhi in 1964 by a notification under the Union Territories (Laws) Act, 1950. Consequently eye-banks were

* *Hindustan Times*, September 9, 1983

built up which issue donor cards to persons who wish to donate their eyes after death. The working of the Bombay Act, however, revealed certain deficiencies. Parliament, therefore, decided to replace the Act.

One of the loopholes in the Act was that it permitted the removal of the eyes of a dead person by any registered medical practitioner practising any system of medicine and recognised as a medical practitioner under any law for the time being in force in the country. The task of removing and preserving the eyes needs a special skill, training and facilities and not any and every medical practitioner is capable of doing it. It was, therefore, considered necessary that only those who qualify as registered medical practitioners under the Indian Medical Council Act, 1956, and have a post-graduate qualification in ophthalmology or those who have adequate experience and training in enucleation procedure should be permitted to enucleate the eyes from the body of a deceased person.

Besides, the Act of 1957 had no provision for the making of rules. It was felt that a statute can be more meaningful when supplemented with detailed rules in regard to the matters governed by the Act. Parliament thus enacted the Eyes (Authority for Use for Therapeutic Purposes) Act, 1982. This Act received the assent of the President on August 7, 1982.

Any person under this Act may—in writing or orally, in the presence of two or more witnesses (at least one of whom is a near relative)—unequivocally authorise before his death, the use of his eyes after his death for therapeutic purposes. When the donor dies, the person lawfully in possession of the dead body shall grant all reasonable facilities to a medical practitioner (ophthalmic) for the removal of the eyes of the deceased for therapeutic purposes. Where the person in possession of the body has reasons to believe that the donor has subsequently revoked his consent, he may refuse facilities for enucleation. Also no removal of eyes shall be permitted if the person required to grant such facility has reason to believe that an inquest may be required to be made in relation to such body in pursuance of the provisions of any law; nor can such permission be given by a person to whom the body has been entrusted only for the purposes of final disposal like cremation, interment, etc.

When a person dies in a hospital or prison and the body is not claimed by any of the near relatives, permission may be given by the person in-charge of the management or control of that institution. This authority cannot be given until the expiry of half an hour from the time of death in cases where there is no facility for cold storage of the body, or two hours from the time of the death in cases where facility for cold storage exists.

If the person empowered to give such authority has reason to believe that any near relation of the deceased is likely to come and claim the body, he cannot give the authority.

Quite often bodies are sent for post-mortem for medico-legal purposes or pathological purposes. In such cases, the person competent to give authority for the removal of the eyes may authorise such removal for therapeutic purposes provided that he had reason to believe that the eyes will not be required for the purpose for which the post-mortem is to be done and also provided that the deceased had not expressed any objection to it before death.

The other Act, i.e., the Ear Drums and Ear Bones (Authority for Use for Therapeutic Purposes) Act 1982, has similar provisions in regard to the use of ears of a person after his death. The provisions in respect of giving consent, persons authorised to facilitate the removal after death, the refusal of permission in certain cases, etc., are the same under this Act as in case of the removal of the eyes under the "Eyes" Act.

The statutes also provide for proper steps for the preservation of the eyes and ears after they have been removed from the body. If proper preservation steps are not taken, the whole purpose behind the law will fail. Another very important aspect of the matter which has been duly taken care of under both the statutes is in regard to protecting those who have acted under the Act in good faith. Thus a doctor (medical practitioner) or a relation or any other person in authority to facilitate the removal of the organ, shall not be liable for anything which is done in good faith or intended to be so done under these statutes. No suit, prosecution, or other legal proceedings shall lie against them. If these people are not adequately protected by law, nobody would take the risk of doing or facilitating the task of enucleation.

34

TRADING IN CORPSE*

And now it is trading in human corpse! According to a news report, skulls and skeletons of human bodies are being sold to the West at fantastically high prices. The bodies are "processed" and shipped. The place from where they are procured and where they are processed is not exactly known. Mostly these bodies are those of the Hindus and are shipped from Calcutta. These could be of Muslims and Christians as well.

The modus operandi seems to be to dig the graves and carry off the bodies. In case of the Hindus, there is a likelihood of half-burnt bodies being immersed into the holy rivers wherefrom the unscrupulous dealers pick them up. The prices, according to a weekly in London are, "half-disarticulated skeleton with skull £ 58 (Rs. 928) fully articulated skeleton £ 140 (Rs. 2,240) and first class skull with teeth £ 50 (Rs. 800)". The buyers of these merchandise are mostly universities but doctors and other people belonging to the medical field also are amongst the buyers.

Human bodies can become an object of use and abuse in a variety of ways. The use of a corpse for gratification of morbid impulses, though not very common is not unknown. "Like all perversions, it is as old as mankind and is mentioned by Herodotus as having occurred in ancient Egypt." Apart

* *Hindustan Times*, October 19, 1983

from that, bodies can be stolen for experimentation, transplantation of certain organs and for trading. The law on the subject, however, is inadequate, unsatisfactory and vague.

Under the common law in England, the body after burial was nobody's property. The result of this was that graves were plundered by unscrupulous people. Lack of legal provision for cadavers for anatomical study and the overcrowded conditions in the London Church Yard made matters worse. Later, however, laws were enacted which made it a criminal offence to remove a body from the grave. Legal provisions were also made for cadavers for anatomical studies.

In India, as in countries like the United States, a corpse is not a property and so stealing of the same is not theft under the provisions of the Indian Penal Code. A reference may be made here of a case decided by the Allahabad High Court in 1903 (*Emperor* v. *Ramadhin*). The applicant was tried by Jury at the Sessions for theft under Section 379 of the Indian Penal Code and sentenced to two years' rigorous imprisonment.

The facts were that an elderly man had sustained injuries at the hands of the appellant which, it was said, caused his death. While the body was lying outside the house of the mother of the deceased, waiting to be taken for the last rites, the appellant, with some other men came, seized the corpse and threw it into the river. This was done with the object of destroying evidence that death was due to the injuries inflicted by the applicant. On these facts the jury unanimously found a verdict of guilty as the body had been taken away without the consent of the mother of the deceased and thus theft was committed.

In the revision application before the Allahabad High Court, however, the conviction was set aside, it was held that there could be no property in human being, whether living or dead, and therefore stealing a corpse is not theft as defined in the Indian Penal Code. An exception however was made by the court, in case of "human bodies, or portions of such, or mummies, preserved in museums or scientific institutions." A mention may here be made of the provisions of Section 297 of the Penal Code under which indignity shown to a human corpse is an offence. The relevant portions of the section read as follows:

"Whoever . . . offers any indignity to any human corpse . . . shall be punished with imprisonment of either description for a term which may extend to one year or with fine, or with both."

In this age when bodies are being commercialised, it is essential that there should be adequate legal provision to protect them against abuse. Corpse could be stolen for profit and experimentation and sold at a premium to countries abroad. Organs could be enucleated for transplantation. The purpose behind meddling with a body could be morbid and immoral as well. It is not only a crime but also a sin to indulge in these practices. Mankind has stooped low enough to agonise and torture the living fellow beings but now a time has come when even the dead are not spared. Though law cannot control morality by suppressing baser human instincts, it can at least prevent the overtly horrid acts by making them punishable.

35

CONDEMNED PRISONERS: RIGHT TO DONATE ORGANS*

There has been a sea-change in penological thinking and theories. The purpose of punishing an offender is no longer punitive alone; nor is he viewed as one "diabolically possessed", deserving the severest treatment. The focus is more on reformation and expiation than on inflicting torture. This is evident from the welfare schemes for prisoners and the sensitivity with which the judiciary views any invasion of their human and fundamental rights. Prison welfare agencies, judicial generosity and public interest litigation have played an important role in this matter. Denial or infringement of the rights of the prisoner, or any other person, may be by way of acts of commission, e.g., infliction of cruelty etc., or by acts of omission, as by preventing him to do something which he has a right to do.

It is interesting to note a recent phenomenon in this direction. Several prisoners facing life sentence or condemned to death wish to donate their organs to atone for their misdeeds. In Gujarat the *Sadavichar Parivar*, a social work organisation, has appealed to the convicts in the Sabarmati jail to donate their kidneys. Similar appeals have been made to prison inmates in other parts of the country and the response

---

* *Hindustan Times*, March 22, 1984

is believed to be encouraging. Ranga, the Chopra children's murderer, had expressed a desire to donate his eyes. The authorities, however, have refused permission to the inmates to donate their organs. A specific instance may be cited here.

Question of Rights

A convict sentenced to death in Rajkot jail wished to donate his kidney to an ailing patient but the State Government declined permission on the ground that it would be against the rules. The rules under the Jail Manuals of almost all the Sates provide that the person to be hanged should be medically examined and found medically fit. The authorities fear that removal of a kidney might endanger the health of the donor convict and thus give rise to problems in carrying out the sentence of execution. The convict has filed a writ in the Gujarat High Court against this refusal. Such a rule and interpretation thereof, gives rise to various issues.

Coming to the question of rights, it would be pertinent to note two articles in the Universal Declaration of Human Rights. Under Article 3, "Everyone has the right to life, liberty, and security of person". Article 29 says: "(1) Everyone has duties to the community in which alone the free and full development of his personality is possible. (2) In the exercise of his rights and freedom everyone shall be subject only to such limitations as are determined by law solely for the purpose of securing due recognition, respect for the rights and freedoms of others and of meeting the just requirements of morality, public order and the general welfare in a democratic society."

In consonance with Article 3 of the Declaration, our Constitution, vide Article 21 lays down as a Fundamental Right that "no person shall be deprived of his life or personal liberty except according to the procedure established by law."

No Iron Curtain

The question here is, does personal liberty include the right of an individual over his body; and secondly, upon conviction does a convict become bereft of his Fundamental Rights including the one to personal liberty?

It is well known that a free person is at liberty to donate his organs—kidneys, eyes and ears etc., and, in fact, it is very much appreciated. There are appeals for such donations. In other words, a person has the liberty to use his organs—donate or gift—as he pleases. When a person under a sentence is denied such permission should one take it to imply that his fundamental right to personal liberty is annihilated in toto?

As pointed out by Justice Douglas of the United States Supreme Court, "There is no iron curtain drawn between the Constitution and the prisons . . . ." Constitutional rights do not desert convicts upon conviction but become "dwindled in scope."

In the words of Justice Krishna Iyer (former Judge of the Supreme Court) in *Sunil Batra's* case (1978) "prison laws do not swallow up the Fundamental Rights of the legally unfree . . . . The policy of the law and the paramountcy of the Constitution are beyond purchase by authoritarians." And again, "Prisoners have enforceable liberties devalued may be but not demonetized." Thus it is argued that the right to donate organs is a personal liberty which is a fundamental and human right of every individual and conviction does not usurp such rights. What imprisonment or conviction may justifiably do is to curtail the liberty and freedom of movement. Imposition of death sentence is also provided in exceptional cases though its legality or propriety is a matter of debate and opinion.

Article 29 of the Declaration provides protection to a person's right to do his duties and to the exercise of his rights and freedoms. Is it not everybody's duty to do whatever good and noble he can for his fellow beings? The rights and freedoms are subject to certain limitations imposed in the interest of securing due recognition and respect for such rights and freedoms of others, and in the interest of morality, public order and general welfare. A convict donating his organs would not, by any stretch of imagination, invade any of these provisions.

What is the criterion or degree of medical fitness that a convict has to have? The Rajkot prisoner was refused permission to donate his kidney because it was feared that his health might be endangered and he may not be medically fit to go to the gallows. Is only physical fitness the criterion or is mental fitness relevant too? If the latter is considered to then a majority of those serving a long sentence or those who are condemned prisoners are mentally shattered, depressed and sick. Besides, does a person who donates his organ become medically unfit? If so, medically unfit for what? A person may be medically fit for one purpose but not so for another. Would a blind or deaf or dumb person be described as medically fit or unfit? If medical or physical fitness is a criterion, the absence of which might save or postpone a person from going to the gallows, is there not a likelihood of prisoners resorting to self-infliction? After all, it is not all that impossible to smuggle or procure offensive weapons. Also would handicapped offenders convicted to death be saved too?

Medical Fitness

Meeting the argument of medical fitness, it is common knowledge that donation of organs does not necessarily make a person medically unfit. There are many who have lived long with one kidney, with one eye, or with damaged ear drums.

A person convicted or hanged for an offence has satisfied the sentence. The question is, does his dead body also continue to be subject to restraints and restrictions to which he was subject when alive? What are the rights of a dead convict? Obviously, he can make a will and specify the course and user of his property. Can't he donate or will his eyes, kidneys or other organs to be used for the benefit of somebody or for medical research? Cornea (eyes), ear drums (ears) and kidneys can be transplanted from live donors as well as from cadaver donors. These organs are of no use to a person who has breathed his last but can give a fresh lease of life to one in need of these. There is nothing in law to hold him from willing away his organs. After his death his wishes should be honoured in the same way as the wishes of any other deceased person. The rights of the living may be curtailed by a procedure established by law, e.g., by way of restraints on movement etc., but not of the dead, especially when there are no risks involved, like disturbance of public peace etc. What risks can be involved in a medical enucleation of organs in consonance with the last wishes of a convict?

Feeling of Atonement

There are about a thousand jails in the country. There are many lifers and some are in the death row. Many of those who languish in guilt and remorse would welcome the idea of donating an organ and thereby doing something good for mankind; the very idea of service would be elating to them—who have inflicted wrong or injury on somebody; it would give them a feeling of atonement, that some sin has been wiped out by the good deed. Moreover, the cost of a kidney is very high and there are many deserving patients who cannot afford to buy it. What is the harm if those who wish to donate—whether they are free or unfree—are given permission to do so? Why be fussy and over-technical about a gesture which is so sublime and noble? Many have killed the living but when someone wishes to save a dying person by parting with his organs, why not assist and appreciate the gesture?

The jail Manual rule which provides for medical fitness probably seeks to ensure that a condemned prisoner should not be ignored or denied medical care simply because his end is imminent; nor can he be encouraged to do something that might endanger his health; but if he, with full knowledge, volition and with the desire to help or save somebody, wishes to do an act which is by all standards and interpretation noble, no technicality should stand in the way.

36

LIVING BEYOND LIFE*

We are living in an age of legalism where everything has a legal connotation, implication and reaction, medicine not excepted. Law and medicine are two disciplines which touch the life of every individual. No man can remain impervious to the developments in these two areas; the degrees of such contact, however, might vary. Until a few decades back the interaction between these two disciplines was rather obscure. Times have changed, and changed very fast, and we find that the two are inextricably connected. The World Congress on Law and Medicine held in Delhi recently (February 22–25) highlighted the close interaction and relation between the two.

It cannot be denied that medical science has made tremendous advancements in various respects—before life, during life, at the time of death and after death. Perhaps medicine is the greatest intruder in the realm of Nature and given a choice, law would have passed a verdict of trespass against it. But no! Medical science has given hope of life to the dying, hope of progeny to the sterile: hope of physical movement to the handicapped and disabled, and much more.

An area which is of great interest and practical importance in recent years is that of organ transplantation. It is common knowledge that many people live incomplete lives because of

---

* *Hindustan Times*, March 17, 1985

impairment of some of their organs like ears, limbs etc. Life of many is cut short because of diseased vital organs like kidney, liver or heart. Medical advancements have brought hope in the lives of these people but not without the assistance of law; for everything needs to be regulated.

Transplantation can be of various kinds viz:

(a) animal to human, known as heterotransplantation;
(b) human to human known as homotransplantation;
(c) transplantation between twins which is called isotransplantation; and
(d) transplantation within the patient himself, known as autotransplantation.

There is still another category viz. transplantation of artificial organs. A transplanted tissue or organ is a foreign body and hence the donee's body is very likely to reject it.

Animal to human transplantation are least successful. Homotransplantation succeed in varying degrees. Categories (c) and (d) viz. isotransplantation and autotransplantation are most successful. So far as artificial organs are concerned, apart from the high cost, it may not be possible to have them for other than relatively simple mechanical functions.

It is interesting to note that organ and tissue transplantation date back to over 5000 years when the Egyptians and Hindus "transplanted skin to replace noses destroyed by syphilis." It is said that in 1760's John Hunter, with the aid of a dentist transplanted teeth for fashionable ladies. Organs of animals were transplanted (within animals). The mid sixties witnessed rapid strides in human transplantations and the first successful heart transplant is believed to have been performed in Johannesburg in 1967. Donation of blood and transfusion of the same to another is very common. Donation of kidneys and other organs is also not unknown.

At first thought it might appear that the whole process is simple enough and that if the donor and the donee agree to the process, law need not enter into the transaction. Things, however are not so simple as they appear and legal issues of varied nature are attracted in organ transplantation from one to the other, involving the donor, the donee and the medical profession.

The first question here is as to who can donate an organ? The answer is, any person of age and sound mind, a minor or incompetent with consent of guardian and even a dead body through the legal heirs. The question of consent is of vital importance. The consent has to be (1) competent, (i.e., the person should be of age and mentally fit to consent); (ii) informed (i.e., the donor should know what the procedure is and the risk and likely effects of the same); and (iii) free (i.e., there should be no pressure, force or coercion on the person agreeing to donate). Absence of consent by the donor, even if he is an adult, is bound to create complications for the physician as happened in *Fleming* v. *Mut. Liability* Co. (1966), an American case, where a donor charged the physician with negligence for removing skin from her inner thigh.

A donor who initially agrees to donate an organ or tissue, but subsequently withdraws, cannot be compelled to fulfil his promise. A reference may here be made of an interesting case, from America again, (*Mcfall* v. *Shimp* 1978). Mcfall, a 39-year old male, suffered from aplastic anemia. His condition required treatment involving a bone marrow transplant. The closer the genetic match, the more likely was the success of the transplant. Mcfall had three brothers and three sisters none of whose bone marrow was sufficiently matching. David Shimp was a 43-years-old first cousin and his bone marrow was found compatible. He agreed to the donation and underwent initial tests to ascertain compatibility. However during the final testing scheduled for a particular date, Shimp did not turn up. Mcfall's suit seeking specific enforcement and compelling Shimp, by a court order, to donate his bone marrow for transplantation, failed as the court said: "in the view of this court, the refusal of the defendant is morally indefensible. (But) for our law to *compel* the defendant to submit to an intrusion of his body would change every concept and principle upon which our society is founded."

In case of those incompetent to give consent because of their minority or because of mental inmaturity, the consent should be given by the legal guardian and there is a heavy burden to prove that such consent was given. The guardian must give consent to such procedure in the interest of the incompetent. In *Bonner* v. *Moran*, (1941) in America a surgeon who operated upon a minor without consent of parent was held liable. There is an interesting case in this context from Kentucky (USA) *Strunk* v. *Strunk* (1969). The Strunks had two sons; Tommy aged 28 years, married, working and a part-time student, suffering from a fatal kidney disease and Jerry, aged 27, incompetent, feeble minded with low IQ and committed to a state institution. Jerry was the only relative found compatible for transplant of his kidney to his brother. The mother, as guardian petitioned the county court for authority for such permission. The court found the operation necessary and felt that since Jerry was emotionally and psychologically dependent on his brother Tommy the only sibling Jerry had, his well-being was weighted against the loss he would suffer if he lost his brother, Tommy. The court consequently permitted the removal and transplantation of the kidney.

Prisoners are as much human beings as others and hence Justice Cardozo's statement "Every human being of adult years and sound mind has the right to determine what shall be done with his sown body", ought to be as much applicable to them as to the free men. Apart from helping those in need and giving them a fresh lease of life, such donation will give the donor a sense of atonement and satisfaction. In this context a reference may be made to a case in India, perhaps the only one, which is still pending in the Gujarat High Court. A condemned prisoner in a jail in that state wished to donate one of his kidneys but the prison authorities have refused him this permission. Their argument is that under the Jails Manual a prisoner who has to be hanged must be in good health and if one kidney of the prisoner were allowed to be removed, the provisions of the Manual would stand violated.

Against this refusal, the prisoner has filed a case in the High Court and the verdict is anxiously awaited. It may be pointed out here that the idea behind the provision of the manual is that the authorities should not become apathetic to the health of the condemned prisoner thinking that since he has to go to the gallows, his health does not need attention. Moreover, removal of an organ would not make him medically unfit or diseased. Suppose, instead of the kidney, the prisoner wanted to donate his eyes or ear drums? Would enucleation of these make him medically unfit? If so, would a prisoner who meets with some accident while in prison and loses an organ be spared of the gallows just because he does not satisfy requirement of the Manual? It is submitted that prisoners should not only be permitted but rather encouraged to make such donation in life and after so that they can serve humanity and atone for their misdeeds. They (condemned prisoners) are at the fag end of their journey; they can by donating their organs add to somebody else's span of life.

Coming to dead bodies or cadavers, a person may during his lifetime make a gift of all or certain of his organs or if that has not been done, the legal heirs or persons legally incharge of the body may consent to enucleation of his organs. It is heartening to note that India has made a beginning by enacting certain state and Central laws especially for donation, in life and after, of eyes and ear drums. The question of legal authority and consent, however, must be precisely set out to avoid legal complications. In *Williams* v. *Hoffman*, a case from Wisconsin, husband alleged that his wife was kept alive for about 24 hours after being declared dead. This was done by physicians through the use of 'life support system', so that her organs could be removed for purpose of transplantation and use.

Thus we see how man can serve mankind in such a wonderful way. What is needed is a spirit and proper education and information. We offer lavish gifts to our near and dear ones but in a material form. Why can't we think of a gift which gives to another much more comfort and joy than a worldly or material gift? Upon death, properties pass on by succession or will. The donor is left with no interest in anything, why can't organs be willed to deserving people or to the state who can select the deserving recipient. Agreed, there is a risk of exploitation, abuse and commercialisation, but this can be sorted out or atleast minimised by proper legal provisions. The donation should be absolutely voluntary without pressures of any sort.

37

ORGAN TRANSPLANTATION: NEED FOR LAW*

The development of science and medical technology has brought about situations not previously contemplated. One such is the transplantation of human organs. There is need to have effective laws to cope with this medico-scientific development including on changing the definition of death. Some countries like Canada, Australia, England and the US already have some laws on the subject. In India also serious thought is being given to it. The need for such a law is being strongly felt because organ transplantation is becoming more common and feasible now and there is pressure to change the definition of death to brain death to facilitate organ transplants from cadavers. But while this is necessary there is need to have strict laws as this is likely to give rise to commercialisation and exploitation. There are already reports of rackets which trade in human organs and corpses. In order to curtail these practices and also to regulate transplantation in the interest of the donor, the donees and the medical profession, it is important to have a law on the subject.

There are certain organs which can be donated during lifetime without much peril to the donor, while some can be used only after the person's death. Eyes, ears and kidneys

---

* *Hindustan Times*, April 13, 1989

come in the, former category; the heart in the latter. The former category organs can also be transplanted soon after the death of the donor. It may, however, be pointed out that the entire process is not devoid of complications giving rise to intriguing issues involving the donor, the donee and the medical profession. Though a central law on organ transplantation is yet awaited in our country, a reference may be made of two recent statutes, viz., the Eyes (Authority for use for Therapeutic Purposes) Act, 1982 and the Ear Drums and Ear Bones (Authority for use for Therapeutic Purposes) Act, 1982. These statutes provide the terms and conditions under which a person may donate his eyes and ears. As early as 1957 the State of Bombay made a law known as The Bombay Corneal Grafting Act, 1957 which made a provision for use of the eyes of deceased persons for therapeutic purpose. The Act was extended to Delhi in 1964. In 1982, however, the 1957 Act was replaced by a new one. Both the statutes relating to enucleation of eyes and ear drums provide the terms and conditions under which a person may donate these organs and also for proper steps for their preservation after they have been removed from the body. The Acts also protect doctors and other persons who have acted in good faith in the matter.

The question of consent is the most crucial in cases of transplantation of tissues or organs. The consent must be competent, i.e., by a person of age and mental maturity, and free. But, a donor who initially offers to donate an organ or tissue but subsequently withdraws his consent cannot be compelled to fulfil his promise. In an American case, *McFall* v. *Shrimp*, 1978 the court held that "to *compel* Shrimp to submit to bodily intrusion would defeat the sanctity of the individual and would impose a rule that would know no limits and one could not imagine where the line would be drawn".

The issue of consent assumes more significance in the case of an incompetent person where consent has to be given by the parent or the guardian or the court. Sometimes the court takes into view the psychological benefit to the donor in the life-giving gift to a seriously ill sibling. Thus in *strunk* v *Strunk* (Kentucky, U.S.A., 1969) the court felt that since feeble minded J was emotionally and psychologically very dependent on T, the latter's death for want of kidney donor would have a very detrimental effect on J. The court, consequently gave permission for removal and transplantation of J's kidney.

As pointed out above, organs can be used for transplantation even from the dead, i.e., cadavers, within a certain period of time. In such cases a person may during his life-time make a gift of all or certain of his organs or, if that has not been done, the legal heirs or persons legally in charge of the body may give consent. The question of legal authority and consent, however, must be precisely set out by legislation to avoid legal complications. The development of artificial, mechanised life sustaining devices like respirators and other bio-medical technologies has made it possible to maintain heart beat and respiration in bodies which have suffered brain death. Thus, we might probably one day see "bioemporiums"" in which there are "corpses. . . utterly unlike any corpses known throughout man's previous history . . . warm, respiring, pulsating, evacuating and excreting bodies, requiring nursing, dietary and general grooming attention. . . long silent rows of them, looking like any other patient would be laid out for the purpose of being, in effect, farmed and harvested for the benefits of the living."

In a Wisconsin case, *Williams* v. *Hoffman*, a husband alleged that his wife was kept alive for about 24 hours after being declared dead. This was allegedly done by the physicians through the use of a life support system so that her organs could be removed for purposes of transplantation and therapeutic use. Thus dead bodies can be specially "farmed" for use of their organs. Time is not far when corpses or dead bodies will be more precious and valuable than a living person as they will be a source of parts for transplant into disabled or ill living beings.

The need of the hour is to have a comprehensive and clear law on the subject. The risk of trading, and commercialisation is very great. There must be effective safeguard against this. The poor must be protected against exploitation of their property. According to one report, "a Briton has proposed to sell a kidney to any needy Arab" for \$ 68,000 to ease his house mortgage repayments" (*Globe* and *Mail*, Toronto, 1977).

The issues of consent must be clearly laid down. No donation should be permissible during life time if it causes a significant risk to the life of the donor no matter how willing he is to make the sacrifice. There are many tissues and organs which can be donated without much peril to the donor but in all cases, the donor should be apprised of the operation, the pain and suffering he may have to undergo and the possible effects, and so on.

Last, but not the least, it would be in the fitness of things that people be educated to will away their organs so as to give a fresh lease of life to the ailing and the dying.

38

SUPPLY OF HUMAN ORGANS FOR TRANSPLANTATION*

Introduction

Medical science and technology have made miraculous advances in prevention and treatment of disease and considerably succeeded in prolonging life, or rather postponing death.

Death is inevitable. It has to come to everybody, sooner or later. It is particularly tragic if the end comes not from a natural process of ageing but from accident or failure of specific organ in comparatively young age.

A remarkable medical technology which has given a hope of fresh lease of life to those afflicted with diseased or defective organs is by replacing those organs with healthy ones.

Transplantation of organs and tissues is not a recent phenomenon. In fact, over 5,000 years ago, the Egyptians and Hindus transplanted skin to replace noses destroyed by syphilis. It is believed that in the 1760s John Hunter, with the aid of a dentist transplanted teeth for fashionable ladies. Transplantation of bones dates back to 1878. However, during the past few decades, the pace of advancement in this field of medicine has been very fast. The first successful corneal graft

* 34 Journal of Indian Law Institute, 399–415 (1992)

in man was performed by Dr. P. Zirm who grafted the cornea of a boy into a man whose eyes sustained burns.

In 1947 a kidney was transplanted with limited success when a kidney was attached to a patient's arm. The first modern heart transplant was performed in 1964 and in the same year six liver transplants were attempted too, though without success. The first allotransplantation of the human lung was performed by Dr. J. Hardy and his associates in 1963.[1] During this period transplants of endocrine glands, pancreas, intestines and bone marrows have also been reported but the rate of success has not been satisfactory because of the rejection of the donor organ in the recipient body. The first major and vital organ to be transplanted successfully was the kidney in 1951 by Dr. David Hume. All these however, remained only experimental procedures until immuno-suppressive drugs and other techniques were discovered which contributed tremendously towards the success of organ and tissue transplants.

Transplantation can be of various kinds, *viz.*,

(a) Animal to human, known as heterotransplantation,
(b) human to human, known as homotransplantation,
(c) transplantation between twins, known as isotransplantation; and
(d) transplantation within the patient himself known as autotransplantation.

Though heterotransplantation has not met with much success so far, yet the possibility of this as a source of organs in future cannot be ruled out. In the first modern human heart transplant which was undertaken in January 1964, the recipient was a 68 year old man with severe heart disease and the donor was an adult chimpanzee.[2] Transplantation of a chimpanzee kidney into a man who survived for eight months before he died of pneumonia is also reported. Possibilities of medical science succeeding in replacing human organs with artificial organs cannot be ruled out either. It may be remote but not unlikely.

The fact however remains that non human supply of organs for use for human transplantation is not imminent and, therefore, for long times to come the demand has to be met from human organs only. The improvement in techniques and technology—the invention of the ventilator and respirator, and the development and refinement of immuno-suppressant drugs—have considerably improved the success rate of organ transplantation. This has created an imbalance between the demand-supply ratio. The number of needy recipients is much more than the donors who can supply organs. Hence the need to discuss methods and alternatives to increase the supply of organs.[2a]

One development which is very significant in this context is the redefinition of death.

Brain Death

The traditional concept of death based on cardio respiratory failure is scientifically inadequate. Under section 46 of the Indian Penal Code, "Death denotes the death of a human being unless the contrary appears from the context."

Section 29(B) of the Registration of Births and Deaths Act 1969 defines death as "the permanent disappearance of all evidence of life at any time after live birth has taken place."

This existing definition of death does not permit removal of human organs from brain dead persons. It is therefore necessary to operationalise the concept of brain stem death and get "heart beating donors." which requires that the organ be removed when the donor's circulation is still intact for best results in organ transplantation technology.

Brain death is a state when a person's brain is so severely damaged that recovery is impossible although in an intensive care unit his heart may be kept beating and his lung working with artificial support.[3]

The WHO has accepted the concept of brain death and advised its member countries to implement these concepts and have legal enactments recognising brain death. Many countries have accepted the brain stem death as the legal definition of death. We too are heading towards it.[4]

The acceptance of brain death criteria, however, poses certain risks, fears and apprehensions. A doctor may certify a person to be dead who is not infact dead, either accidentally or purposely. Reference may be made of a discussion on brain death on a popular BBC programme *Panorama*. It was said that the existing criteria of brain death need not be accurate. Persons who were declared dead were in fact alive and one person who had been declared brain dead a year before walked into the studio. On scrutiny it was realised that all these people had been declared dead by the misapplication of the criteria for death.[5]

The acceptance of brain death as the definition of death also gives rise to an apprehension that it will lead to premature removal of organs. The people will assume that the concept of brain death is being legalised only to give legal permission to remove their organs. A horrifying incident was reported recently in American newspapers. A young boy diagnosed to be dead using some dubious criteria was laid on an operation table. His chest was open and his heart was about to be removed and put into the chest of a man in an adjoining theatre. The police arrived and stopped the operation. The boy's chest was sewn and he is alive and well.[6]

Nonetheless, it cannot be denied that the recognition of brain death has many advantages too. It saves the patient and the relatives of a lot of agony and pain. When a patient is irreversibly ill, sustaining him on artificial respirators and other devices would be merely prolonging the process of death. Besides, it would mean waste of money and scarce hospital resources which could rather be utilised to save the lives of those who are not hopelessly ill, though emotionally speaking, this argument sounds crude.

It is, however, very essential to provide a clear-cut definition with proper safeguards. The criteria and rules for diagnosing brain death should be clear and uniform so that complications do not arise. A reference may be made of one incident narrated by a doctor during a workshop on brain death, held at Calcutta in 1990. A brain dead patient was taken to a well known hospital in South India and put on a respirator. Her son had come from USA. He told the doctor, "back home this person would be declared dead;" so the doctor said, "I agree with you. Do you want us to turn off the respirator?" The son replied in the affirmative, signed a form and the mother was taken away. A month later, the second son came to the doctor and complained, "Doctor why did you stop the respirator when my mother was not dead. We in India, do not recognise brain death. If my mother had been alive today, I, and not my brother, would have inherited all her property."[7]

Thus vagueness or uncertainty in the definition of death could lead to complications and problems.

Sources of Organ Supply

The source of organs supply could be cadavers or live donors. Both involve various issues and implications but in case of live donors, the complexities and apprehensions are much more than in case of cadaver donations.

Cadavers

Even before medical science reached its present stage when cadaver organs can be used for saving human life, cadavers have been of great value for traditional medical purposes like autopsy, study of anatomy, research and experimentation. According to one report in the late eighteenth century, when the medical need for bodies for use in medical schools was greater than the supply, cadavers were purchased because there was no other way to procure them. This gave rise to professional grave robbers called "resurrectionists." Grave robbing and nefarious activities of Burke and Hare, who murdered people to supply bodies to Dr. Knox, well known Edinburg anatomist, led to the enactment of a statute in England.[8] Under this Act medical schools were supplied with corpses of paupers and unclaimed dead. Most American states followed suit and enacted similar laws.

The value of cadavers has now gone up all the more because of the use of its parts for transplantation. It is, however, a matter of compunction that "tens of thousands of individuals are suffering and dying while the organs that could restore them to health are disposed of like carrion."[9]

The need of the time is to have proper laws and regulations for salvaging cadaver organs.

Bequests or express consent

The simplest and most common way to procure organs is by permitting people to make bequests of their organs. Many countries have made such provisions. In 1947, California enacted laws regulating the bequests of organs. A citizen was given a right to donate his organs under a will and heirs and executors were under an obligation to honour his wishes. Most other states also enacted similar laws. Today, the major source of organs in the USA is donations made at donor's death. The consent for the same is given either by the donor during his lifetime or posthumously by his family. The Uniform Anatomical Gift Act (UAGA) 1968, regulates these donations. Any adult of sound mind can either permit or forbid the use of his organs after his death for purposes of research, teaching or tranplantation. The donation may be made of the entire body or part of it by will or by a non-testamentary document like a donor card. In Colorado, those applying for a driver's licence have to specify whether or not they wish to permit use of their organs in the event of their death. If the individual fails to express his preference, the next of kin can donate his organs posthumously.[10]

Potential donees include organ banks, hospitals and doctors, accredited medical schools and individuals in need of transplantation.

It is important to note that the Act is purposely silent on the issue of payments for anatomical gifts. Prior to the Act, a few sates prohibited the sale of human organs and bodies after death. They banned an individual from arranging to sell his organs after death but did not expressly ban the posthumous sale of his organs by his next of kin. In 1984, the National Organ Transplant Act was passed. It prohibits organ sales and imposes heavy penalties on those who violate this provision. The immediate cause that led to the enactment of this statute was Dr. B. Barry Jacob's proposal to buy and sell human organs. He set up a Virginia company to broker human kidneys. Unlike an earlier proposal by another entrepreneur under which people could sell their organs after death with the proceeds to go to their beneficiaries, Jacobs solicited kidneys from healthy live donors. He proposed to charge £ 2,000 to £ 5,000 as his brokerage. He also planned to attract poor people from Third World who could sell their kidney for nominal price.

The English Human Tissues Act 1961, which regulates the use of parts of bodies of deceased for, *inter alia*, therapeutic purposes, also provides for bequests.

Section 1(1) of the Act says:

> If any person, either in writing at any time or orally in the presence of two or more witnesses during his last illness, has expressed a request that his body be used after his death for therapeutic purposes or for purposes of medical education or research, the person lawfully in possession of his body after his death may, unless he has reason to believe that the request was subsequently withdrawn, authorise the removal from the body of any part or, as the case may be, the specified part, for use in accordance with the request.

This Act was silent on the aspect of payment for organs. In 1989, the Human Organs Transplants Act was passed which clearly prohibits all commercial dealings in human organs removed from dead or living persons. It also imposes restrictions on transplants of organs removed from live donors between persons not genetically related.

Presumed consent

Another suggestion to increase the supply of organs is to permit removal of organs as a matter of routine unless the decedent had "opted out," *i.e.*, expressed his objection during his lifetime, or there is an objection by the next of his kin, after his death. As remarked by Professor Dukeminier:[11]

> [B]y making the basic presumption one which favours life, and by thus putting the burden of objecting upon persons who would deny life to another, the policy of saving human life is given first priority, and the wishes of persons to preserve a corpse inviolate are also accommodated. This method would produce far more organs for transplantation than are produced by statutes permitting organ donation by the decedent.

This proposal, however, is also not likely to be very popular. It would be very difficult to ascertain whether or not there was an objection. Even though the presumption in such cases would be against any objection — yet the surgeon would not want to remove the organs for fear of any complexities.

Consent of Relatives

Yet another alternative suggested for procurement of organs for transplantation is by requesting the relations of the deceased for the donation. This was first proposed in 1984 by Arthur Caplan.[12] A physician would ask the next of kin or request them to donate the organs of the deceased. This is however, an extremely delicate task. Approaching the family of the victim whose life has just ended, and asking them to donate his organs would be an act of extreme insensitivity to the bereaved families' state of mind and utterly embarrassing. The family, who is in a state of trauma and shock would be too hurt with such a request. As narrated by a lady who was requested to donate her husband's kidney:[13]

In my state of acute shock, distress and grief, there suddenly came this totally unexpected question—I was astounded and utterly appalled at such a complete lack of feeling. To make such a decision for oneself is hard enough but to be asked to make it on behalf of another while one is so shocked and grief-stricken, is both harrowing and cruel . . . . Never could I want any close relative to suffer as I had done in making such an agonizing decision during the worst moment of a life time.

In another incident narrated by a doctor at a seminar on brain death in Bombay recently the relatives of a brain dead person were standing nearby when a nephrologist approached them for kidney donation. One of the relatives said, "doctor you go away before I take out my revolver and you become brain-dead."[14]

Compulsory Removal Irrespective of Consent

Another alternative suggested to meet the growing demand of organs is to have a legislation authorising removal of cadaver organs irrespective of consent or even regardless of objection by the deceased or his kin. This mode of compulsory removal of cadaver organs is sought to be justified on the plea that saving life is paramount to all other principles. It is argued that all other considerations should be subservient to this. Such a procedure, however, is bound to be assailed on various fronts. It might be considered to be an act of indignity to the corpse to commit trespass on it and mutilate it for removing organs. Also, certain religions might consider removal of organs to be inconsistent with their tenets which enjoin that the dead be buried intact. For instance, to a fundamentalist Christian it might be inconsistent with the principle of bodily resurrection. Likewise shedding of blood might be objected to by a Jehovah's witness. Many orthodox rabbis have opposed autopsies involving a principle of Judaism that the body must not be violated.[15]

Inhibitions in Cadaver Donations

Apart from what has been said above there are various other factors which might inhibit a person from donating cadaver organs.

Psychological

No one wants to think about his death. Signing donation papers, therefore, arouses deep psychological problems, tensions and anxieties. As remarked by Freud:[16]

> Our own death is indeed unimaginable and whenever we make the attempt to imagine it we can perceive that we really survive as spectators. Hence . . . at bottom no one believes in his own death, or to put the same thing in another way, in the unconscious every one of us is convinced of his own immortality.

According to some studies, one of the primary reasons why people do not make wills of their property is their difficulty in facing death. If that be so, donating organs is still more difficult. "Organs are so much a part of a person's conception of himself that signing a donation paper usually arouses the deepest and most fearful anxieties."[17] It also gives rise to apprehensions that the best may not be done to treat him and save him from death.

Religious

Some religions have expressed mandates against the disfiguration of the dead body. Autopsies are "abominations of the body" according to the orthodox Jewish, and to the Christians, organ removal is inconsistent with the principle of bodily resurrection. The most pervasive belief is that proper burial displays respect for the departed person. The body has to be shown all respect because it is spiritually imbued.

The history of cadaver disposition has been greatly influenced by the superstitions that surround the dead body. In England, from the time of the Roman conquest to the nineteenth century, the right of disposing a body was solely within ecclesiastical cognisance, perhaps because of the Church's concern for prevention of sacrilege. "The Church took the body to itself. It held that a corpse was appropriated by it, by divine service and consecrated burial. The spirit departed to the realms of the super-natural; the body was held by the divine agent to await resurrection."[18]

Live Donors

Non-vital or paired organs can be removed from a live person without much of peril to him. For example, the removal of a kidney, spleen, bone marrow, part of thyroid, cornea, skin, ear drum etc. is well known. Live donations, however, give rise to various social, legal and ethical issues. Many countries put a statutory ban on removal of organs if it could produce a permanent deficiency.[19] Under the common law an adult, who is fully informed of the risks and consequences, may part with his organs which do not involve risk of death or impairment of bodily functions. Determining consent, however, is not easy and simple. Various extraneous factors like familial pressures, economic coercion or unfair inducements might impel or motivate a person to give consent. Such consent may not essentially represent free will. There are bound to be difficulties especially in the case of minors and the mentally weak.

Another basic principle which is invoked against live donations is that it is an offence to commit a trespass into another's body by even touching it without the consent of that person or without a proper legal justification unless done for that person's benefit. Mere pecuniary benefit is not considered to be a benefit under any system of law, in this matter.[20]

Besides, it is also an offence to cause injury or grevious hurt to a person even if he consents to it. Removal of an organ could constitute an act of grievious hurt.

Whatever be the basis or arguments against live organ donations, the fact is that a complete ban will never work. It would rather give rise to underhand practice, toutism, exploitation and black marketing. What is needed therefore, is a regulation of such donations with proper safeguards. Live organ donations should be permitted only for transplantation into related persons. When an exception is to be made, there has to be a proper scrutiny by an authorised body to ensure that there are no extraneous factors and that the donor is not giving his organ under any compulsion, pressure or impelling circumstances.

Sales of Body Parts

The sale of body substances, especially blood, is very common. However, when it comes to the subject of sale of human body parts like kidney, liver, eyes, bone marrow, the immediate reaction is one of extreme disapproval and repulsion.

The sale of organs is strongly opposed as an unethical practice since it is feared that the poor and powerless alone will sell their parts and the rich will buy. The poor who need money will sell their organs but the poor who need the organs will die because of their incapacity to pay and buy.

There are also risks of "unsavoury trafficking" if human body parts are made saleable commodities. It might lead to murders or illegal disinterment of corpses for sale[21] or even suicides by the poor so that their families could benefit from the sale of their organs.

As observed by one author;[22]

> There is real danger that unless something is done to improve the efficacy of the voluntary system, advocates of a free market solution will attempt to create a for-profit system to meet the large demand for organs.

His hostility is so strong that he implies that "a market solution would be a cure worse than the disease of thousands of individuals suffering and dying for the lack of an adequate supply of transplant organs."[22a]

Another argument against the sale of organs is that the market system will discourage donations and there will be an "overall decline of altruism in favour of materiality." Drawing support form commercialisation of blood, R Titmuss remarks:[23]

> The commercialization of blood and donor relationships represses the expression of altruism, erodes the sense of community, lowers scientific standards, limits both personal and professional freedoms, sanctions the making of profits in hospitals and clinical laboratories, legalizes hostility between doctor and patient, subjects critical areas of medicine to the laws of the market place, places immense social costs on those least able to bear them — the poor, the sick and the inept — increases the danger of unethical behaviour in various sectors of medical science and practice, and results in a situation in which proportionately more and more blood is supplied by the poor, the unskilled, the unemployed, Negroes and other low income groups and categories of exploited human populations of high blood yielders.

Besides, it is apprehended that if a price is attached to organs, there would be a high risk of diseased organs coming to the market. During a medical examination or at the time of taking the medical history, donors will be tempted to conceal the existence, past or present, of any malady or disease. This has been the experience in case of blood donations. Voluntary donors are less likely to conceal facts and true medical history, like illness, disease, infections, innoculations, personal habits and way of life etc. than paid donors. The payment involved in blood donations is not very high whereas in case of organs the monetary gain is much more and hence the temptation to conceal any fact that might disqualify a person as a donor is higher too. Besides, it is easier to conceal an adverse fact since it is difficult to pin-point the cause for failure of the transplant.

The sale of body parts could also give rise to some queer issues like—can a donor mortgage his organ against a loan? Or can a person be forced to sell his organ to satisfy a debt recovery decree? A reference may be made of a case in America where a judge in Lexington, Kentucky, gave traffic violators option to pay fines in blood.[24] It cannot, however, be denied that creating a market for organs is likely to be the most effective method of procuring sufficient supply of organs which could save innumerable lives. As rightly observed by one writer.[25]

> Although commodification of organs would initially shock a society . . . allowing payment for organs would help eliminate the tragic shortage in transplantable organs.

People who do not want to donate, or those who have otherwise no religious or other inhibitions but are simply indifferent or callous and do not take initiative in the matter, would come forward if financial or other incentives are provided for donations of their organs after death. To quote Lloyd R. Cohen:[26]

> A futures market is a robust solution to the organ shortage problem. Not only does it directly attack the problem of insufficient incentives for people to donate and for their next of kin to acquiesce, it also indirectly and powerfully attacks the more pressing problem of providing incentives for hospitals and medical professionals to facilitate the process. It is like clearing a clogged pipe with a constant high pressure blast of water rather than replacing the suspected clogged section. Wherever the blockage, now or in the future, the financial benefit will flush it out.

Besides, permitting the destruction of usable cadaver organs puts a heavy economic burden on the society. An economist calculates the economic losses on the following basis.

(i) the economic cost of maintaining persons on artificial machines or expensive drugs, and

(ii) the economic and social cost of permitting a person to die simply because of non-availability of an organ.

Hence, it is argued that there is nothing wrong if a price is given for an organ supply.

True, an ideal state would be one where people are moved by altruistic motives to donate their organs but that need not be an argument against complete ban on sale. As very aptly remarked by one author,[27] "charitable donation of organs not only remains possible but in fact becomes a more noble act when sale is permitted . . . . *It is hardly an act of great generosity to donate that which you cannot use and may not sell.*"

The Indian Position

In 1982, the Eyes (Authority for Use for Therapeutic Purposes) Act and Ear Drums and Ear Bones (Authority for use for Therapeutic Purposes) Act were passed. As early as 1957 the State of Bombay had enacted a law known as the Bombay Corneal Grafting Act which was extended to Delhi in 1964. This Act has been replaced by the above mentioned Eyes Act of 1982. Both the statutes relating to enucleation of eyes and ear drums provide the terms and conditions under which a person may donate these organs and for proper steps for their preservation after they have been removed. In 1982, the Maharashtra Government enacted the Maharashtra Kidney Transplantation Act, which regulates the process of removal, preservation and transplantation of kidneys. There is however, no comprehensive legislation regulating the removal of organs from live or dead persons. The large demand and scarce supply of transplantable organs have encouraged rackets and touts who are exploiting the situation and luring the poor, gullible and innocent people into parting with their organs for a nominal price. There are organised rackets which do this business of procuring and supplying organs. Very recently, there was a report that a drug addict was coerced and lured to sell away his kidney for Rs. 30,000 under the influence of drugs. There are innumerable cases where people have sold their organs and tissues to pay off loans, construct a house, set up a small business, meet medical expenses of an ailing family member, and so on.

According to one report, "with rates being fixed for various human body parts, it is beginning to look more like a meat bazar. . . [With rates around] live donor kidney Rs. 30,000, live cornea Rs. 80,000, skin Rs. 1,000 a patch, cadaver Rs. 6,000 [and a] full skeleton, Rs. 10,000"[28] Villivakam, a suburb on the outskirts of Madras, is inhabited by residents who can hardly afford two meals of boiled rice every day. They sell their organs, specially kidneys, to meet their financial needs. Many have sold their kidneys either to pay off a loan, or to buy a bicycle or autorickshaw, or to set up a small shop. The deal is struck through middlemen who take the donors for necessary tests at their own expense and keep a list of their blood group etc. When the deal is finalised, they take anything between Rs. 1,000 and Rs. 5,000 as their fee. In some cases both the husband and wife have sold off their kidneys. In fact, "the suburb is derisively referred to [as] "kidney' vakkam; or more accurately 'Kidney sold' um."[29]

Though the sale of kidneys is common, sale of other organs like the cornea and skin are not unknown either. Touts arrange for a cornea donor who is admitted to a hospital under the pretext of some eye ailment. The cornea is then removed and transplanted. The price for a cornea is around Rs. 80,000. Dr. O.P. Kulshreshta, Chief Opthalmologist of an eye hospital at Jaipur, was reportedly stunned when one of his patients—a young man in his 20's—told him that he was willing to sell one of his eyes for Rs. 1 lakh and wanted to know in which hospital he could do that.[30]

Though trading in organs sounds morally reprehensible the doctors feel that they are helpless. "What does one do when there are no organs forthcoming from cadavers. Do I turn my patient away and say, prepare for your death?. . . As long as it is properly regulated there is nothing wrong with organ transplants."[31] Says Dr. Ashok Kriplani, a nephrologist from Bombay, "What is really needed is finding out ways and means to increase cadaver organs. The need for live donations will obviously come down if sufficient number of cadavers are available and organs procured at the earliest for successful transplantation."

It is heartening to note that a law to curb exploitation and regulate organ transplantation is in the offing. The Transplantation of Human Organs Bill 1992, as introduced in the Rajya Sabha in August 1992, seeks "to provide for the regulation of removal, storage and transplantation of human organs for therapeutic purposes and for the prevention of commercial dealings in human organs and for matters connected therewith or incidental thereto." This law, however, is not to be an all-India legislation as it is to apply, to begin with, only to the States of Goa, Himachal Pradesh and Maharashtra.[32]

The Bill provides for removal of organs from living persons as well as cadavers. A person may, in writing and in the presence of two or more witnesses, give an express consent for the removal of his organs after his death.[33] In that case, the person, who is lawfully in possession of the body, has to provide all reasonable facilities to a registered medical practitioner for removal of organs. The relative may, however, refuse such permission if he has reason to believe that the donor had subsequently revoked his consent. The consent for removal and the removal of organs must be for therapeutic purposes only.[34]

In the absence of any explicit consent or authority for removal of organs by the deceased himself, the person in lawful possession of the body may authorise removal of the organs[35] provided no objection was also expressed by the deceased to such removal, unless he has reason to believe that any near relative[36] of the deceased has objection.

The Bill also provides for removal of organs from bodies lying in hospital or prison and unclaimed by any near relative within 48 hours from the time of death.[37] Such authority is given to the person incharge of the management or control of the institution. No such authority may, however, be given if the person incharge has reason to believe that any near relative of the person is likely to claim the body even though nobody has come forward within the required time, *i.e.*, 48 hours.

Only a registered medical practitioner can remove the organs at a registered hospital after being satisfied by a personal examination of the body that life is extinct.

In order to make organ transplantation technology successful, it is essential that organs should be retrieved before the degenerative process begins. This is one of the reasons why most of the developed countries have recognised and accepted the concept of brain death. The Transplantation of Human Organs Bill also recognises brain stem death. It has been defined as "the stage at which all functions of the brain-stem have permanently and irreversibly ceased."[38]

However, where any organ is to be removed from the body of a brain-stem-dead person, such death has to be certified by a board of medical experts. The board shall consist of the registered medical practitioner who would nominate an independent specialist, registered medical practitioner, and a neurologist or neuro-surgeon from the panel of names approved by the appropriate authority. Besides these three the registered medical practitioner treating the patient has also to be there on the board.[39] The Bill provides for the removal of organs from live donors as well, though restrictions have been imposed on such donations. Cl.9(1) says, "Save as otherwise provided . . . no human organ removed from the body of a donor before his death shall be transplanted into a recipient unless the donor is a near relative of the recipient."

There are, however, some exceptions to this general rule, *viz.*,

(a) the donor wants to give the organ to someone unrelated because of his affection or attachment for the recipient, or
(b) there are some other special reasons for such donation.

In such cases the prior approval of the authorisation committee is required for removing the organ. The donor and the recipient have to make a joint application before this committee and the permission is granted only after the committee is satisfied that the requirements of the Act and the rules have been complied with. The idea behind this provision is to safeguard the interest of the donor and ensure that he is in no way being exploited, coerced or pressurised to part with his organ.

Penalties are provided for removal of organs without authority.[40] The punishment prescribed for such offence is imprisonment for a term which may extend to five years and fine which may extend to Rs. 10,000. In case of an offending registered medical practitioner, his name shall be reported to the concerned medical council for necessary action. Such action may include removal of his name from the register of the Medical Council for two years. If the offence is repeated, his name could be permanently struck down.

In case of commercial dealings in human organs, a minimum sentence of two years imprisonment and Rs. 10,000 fine is prescribed which may be extended to seven years imprisonment and Rs. 20,000 fine. Clause 19 enumerates various dealings which would amount to trading in organs. These are:

(a) Making or receiving any payment for the supply of, or for an offer to supply, any human organ;
(b) seeking to find a person willing to supply for payment any human organ;
(c) offering to supply any human organ for payment;
(d) initiating or negotiating any arrangement involving the making of any payment for the supply of, or for an offer to supply, any human organ;
(e) taking part in the management or control of a body of persons, whether a society, firm or company, whose activities consist of or include the initiation or negotiation of any arrangement referred to in clause (d); or
(f) publishing or distributing or causing to be published or distributed any advertisement,
    (a) inviting persons to supply for payment of any human organ;
    (b) offering to supply any human organ for payment; or
    (c) indicating that the advertiser is willing to initiate or negotiate any arrangement referred to in clause (d).

The Bill also provides for the setting up of appropriate authorities[11] at the central and state levels to look into various matters like granting, renewing, suspending and cancelling registration of hospitals, enforcing such standards as may be prescribed, for hospitals engaged in removal, storage or transplantation of human organs, investigating complaint, inspecting hospitals periodically for examining the quality of transplantation and the follow up medical care to recipients and donors of organs.

The law, however, is still under gestation.*

Conclusion

The foregoing discussion lead to the conclusion that the most viable and expedient alternative to the solution of organ shortage is to provide incentives to people to donate cadaver organs—whether it is in cash or in kind like free medical treatment, rebate in tax, organ priority to a family member should a need arise, and other concessions. Ideally the person himself should enter into a contract or specify his consent for removal of his organs after death but in case he fails to do that, nor does he express any clear objection to it, the next of kin should have the right to give the organs.

Though the relatives may be reluctant to profit from the death of a near one, this compunction can be got over by donating the proceeds of the sale of organs for some charitable purpose like maintaining a hospital or old-age home, or institutions for the poor, etc.

To conclude, any system that alleviates human suffering and death should be acceptable. "When goods are worth more to one person than another, and there are no substantial externalities, the market is a wonderfully efficient mechanism to induce transfer from the latter to the former."[42]

If sufficient number of cadaver organs become available, the need for live donations—with all its apprehensions like danger to life and health of donor, exploitation of poor, commercialisation, inducements by brokers and middlemen etc. — would be eliminated.

There is need for an effective law to regulate donations. An appeal should be made to the people to donate organs. "The point must be made that a death, resulting from unavailability of an organ is neither inevitable nor must it be viewed simply as a statistical occurrence. It must be seen for what it is, a senseless tragedy which could be avoided by overcoming

---

* The gestation is over and it has become an Act in 1994

needless restrictive taboos. At the same time, careful attention must be given to measures aimed at preventing abuse and assuring maximum public acceptability of organ removal.[43]

It would indeed be an act of highest religion to give life to a dying person by donating organs which are no longer of any use for the donor. By doing this, one can live beyond life.

REFERENCES

1. See note "Sale of Human Body Parts" 72 *Mich. L. Rev.*1181–1264, f.n. 6, 1974.
2. Ibid.
2a. The issue of organ allocation for tranplantation has not been discussed in this paper.
3. S. Nundy in *Brain Death and Organ Transplantation in India* (Proceedings of a workshop held at Calcutta on 10 June, 1990, compiled by G.K. Pande, P.K. Patnaik, S. Gupta and P. Sahni) 1, (1990).
4. Transplantation of Human Organs Bill, 1992, cl. 2(d).
5. Nundy, *supra* note 3 at 6.
6. Id. at 19.
7. Id. at 16.
8. English Anatomy Act, 1832, 2 & 3 Will, 4, c. 75.
9. Lloyd R. Cohen, "Increasing the Supply of Transplant Organs," 58 *The George Wash L. Rev.* 4 (1989–90).
10. Colo Rev. Stat. s. 42-2-106(5) 1981.
11. Dukeminier, "Supplying Organs for Transplantation", 68 *Mich. L., Rev,* 811 at 832 (1970)
12. Caplan, "Organs Procurement: Its Not in the Cards" 14 *Hastings Centre Rep.* 9 (1984).
13. Demetrius, "Transplantation: The Relatives' View," *J. Med. Ethics* 71 (1975).
14 As narrated by Dr. Vas in *supra* note 3 at 45.
15. *Supra* note 11 at 826.
16. S. Freud, "Our Attitude Towards Death," in 4 *Collected papers* 304 (1925).
17. *Supra* note 11 at 830.
18. P. Jackson, *The Law of Cadavers* 126 (1950).
19. Italy passed such a law when in the 1930s a rich man bought a testis from a young Neapolitan and had it transplanted. In 1967, the law was modified to permit removal of kidney for transplantation. Sale was prohibited. France indirectly prohibits removal of organs unless it is for benefit of patient. In South Africa, before a tissue is removed, two medical practitioners, other than the performing surgeon, have to certify in writing that no prejudice would be caused.
20. Donation of an organ, however, can be interpreted as an act providing spiritual or psychological benefit and hence for the person's benefits,

for instance, donation of an organ to save a loved one whose death would have caused great shock and trauma.

21. See *supra* note 8.
22. Caplan, "Sounding Board: Ethical and Policy Issues in the Procurement of Cadaver Organs for Transplantation," in *Human Organ Transplantation* 272, at 275, (1987), reference taken from Loyd R. Cohen, "Increasing the Supply of Transplant Organs", *Supra* note 9 at 23.

22a. Ibid.

23. R. Titmuss, *The Gift Relationship* 245–46 (1971).
24. *Time* 113, October 1, 1973.
25. See note "Developments—Medical Technology and the Law" 103 *Harv. Law Rev.* 1625 (1990).
26. *Supra* note 22, at 34.
27. Id. at 28, emphasis added.
28. "Organ bazar", *India Today* 60, 31 July, 1990.
29. Id. at 62.
30. Id. at 61.
31. Id. at 66.
32. The draft legislation was circulated among the states and all, except Himachal and Goa, returned it. Maharashtra already has a law on kidney transplantation which will be replaced by the present law when enacted.
33. At least one of them has to be a near relative. cl. 3(2).
34. "Therapeutic purposes' means the systematic treatment of any disease or the measures to improve health according to any particular method or modality. Cl. 2(0).
35. No such authority can be given if under the provisions of any law an inquest may be required or, if the body has been entrusted to a person only for interment, cremation or, other disposal.
36. Near relative means spouse, son, daughter, father, mother, brother or sister, Cl. 2(i).
37. Cl. (2).
38. Cl. 2*(d)*.
39. Cl. 3(6).
40. Cl. 18(1) and (2).
41. Cl. 13.
42. Cohen, *supra* note 11 at 25.
43. Se note "Compulsory Removal of Cadaver Organs", 69 *Col. L. Rev.* 705 (1969).

39

ORGAN DONATION: CURBING THE RACKET*

The advancements in medical technology are a boon for mankind, but there is scope for their misuse. The ingenuity of the human mind does not take long to exploit these technologies for personal gains.

The technology of replacing diseased and defective organs with healthy ones has given hope of a fresh lease of life to many. The refinement in the techniques and technology has considerably improved the success rate of organ transplantations. This, in turn, has created an imbalance in the demand-supply ratio. The large demand and scarce supply of transplantable organs have made it easy for touts to exploit the situation.

There are reports of drug addicts being induced to sell away their kidneys for a paltry sum. People have sold their organs and tissues—including patches of skin—to pay off loans, construct a house, set up small business or meet medical expenses of an ailing family member. While the sale fo kidneys is most common, trade in other organs is not unknown either. Touts arrange for a cornea donor who is admitted to a hospital under the pretext of some eye ailment. The price for a cornea is around Rs. 80,000.

**Hindustan Times*, March 31, 1994

The chief ophthalmologist of an eye hospital at Jaipur was stunned when one of his patients, a young man in his twenties, told him that he was willing to sell one of his eyes for rupees one lakh and wanted to know which hospital could help him.

Most reprehensible, however, are cases where healthy organs of patients are removed during surgery without their knowledge. According to a very recent report in newspapers, doctors in some hospitals in Orissa are indulging in the lucrative sale of organs.

No law can effectively curb the practice, for there always will be givers and takers and middlemen to facilitate the deal. The only meaningful and practical solution to the problem lies in increasing the supply to meet the demand.

The best way to procure organs is by retrieving them from cadavers. Many countries have enacted laws under which people may donate their organs by will or by a non-testamentary document like a donor card. In Colorado (USA), those applying for a driver's licence have to specify whether or not they wish to permit the use of their organs in the event of their death. Potential donees include organ banks, hospitals and doctors, medical schools and individuals in need of transplants.

Another way of augmenting the supply of organs suggested by some authorities is by removal of organs as a matter of routine unless the person had "opted out" i.e. expressed objection during his lifetime or there is an objection by next of kin. A difficulty which could arise in such cases is of ascertaining whether or not there was an objection. Even if there is a presumption against any objection, the surgeons would be reluctant to remove the organs for fear of being dragged into controversial situations. Another alternative mode could be by requesting relations of the deceased. This, however, again is a very delicate task.

Parliament has introduced a Bill known as the Transplantation of Human Organs Bill, 1992, which seeks to "provide for the regulation of removal, storage and transplantation of human organs for therapeutic purposes and for prevention of commercial dealings in human organs . . . ". It has yet to become a law.* However, even if it is enacted, its efficacy cannot be guaranteed. The need of the hour is to touch the inner-most chords of human hearts and sensibilities by making effective appeals to people to donate organs after death. A thought could be given to provide incentives — whether in cash or in kind — like free medical care, rebate in tax, organ priority to a family member should a need arise, and so on.

Ideally, the person himself should specify his consent during lifetime but in case he fails, without at the same time expressing any clear objection to it, the next of kin should have the right to give the organs. Though the relatives may be reluctant to profit from the death of a dear one, this compunction can be got over by donating the proceeds of the sale for some charitable cause.

* It has now become a law

40

VAST AND VAGUE: TRANSPLANTATION OF HUMAN ORGANS ACT, 1994*

After heated debates and long gestation, the Transplantation of Human Organs Act, 1994, became law in February 1995. Advancements in medical science and technology have made retrieval of organs from dead and live persons and their transplantation into bodies of other needy patients, a possibility. While the improvement in the techniques and technology have considerably improved the success rate of these procedures, it has also created an imbalance in the demand supply ratio.

The absence of a comprehensive uniform legislation regulating the removal of organs encouraged rackets and touts exploiting the poor, the gullible and the innocent into parting with their organs for a price. There have been shocking revelations of "organ-markets" with rates fixed for various parts and tissues; and reports of organs being removed from blood donors without their consent and knowledge by putting them under anaesthesia. There are also reports of donors being "enticed" to go abroad on "kidney-tours" for the removal of commercially viable organs. Hence, the Act.

The Act seeks to provide for the regulation of removal, storage and transplantation of human organs — live and ca-

---

* *Hindustan Times*, July 10, 1995

daver — for therapeutic purposes and the prevention of commercial dealings. A person may, in writing and in the presence of two or more witness, give an express consent for removal of his organs after his death. In case of bodies lying in hospital or prison and unclaimed by near relatives within 48 hours, the person in charge of the management of the institution may authorise removal of organs. Only a registered medical practitioner is authorised to remove organs at a registered hospital after being satisfied that life is extinct.

Since it is important that the organs be retrieved before the degenerative process begins, the Act recognises the concept of brain-stem death which means "the stage at which all functions of the brain-stem have permanently and irreversibly ceased". Such death has to be certified by a board of medical experts.

The Act also requires that the donor and recipient be near relatives which means — spouse,* children, parents and siblings. There are however exceptions such as, when the donor wishes to give the organ to someone unrelated because of his affection; or there are some other special reasons for such donation. In such cases the prior approval of the authorisation committee is required.

Trading in organs is banned under the law which includes making or receiving any payment for the supply of an organ, seeking to find a donor, offering to supply an organ for payment, initiating or negotiating any such deal, publishing or distribution, or causing any advertisement to be published or distributed inviting donors on payment, etc. Heavy penalties are provided for breach of provisions of the Act. There are also provisions for setting up of several authorities to regulate and monitor the working of the Act.

The efficacy of the law in curbing commercialisation is yet to be assessed. It might encourage clandestine practice without proper medical safeguards and also lead to further exploitation of the donor and donee by the middle-man. The seller-donor in financial distress might agree to receive any amount

---

* Apart from being a non genetic linkage, a wife is easily susceptible to pressure, coercion or even duped.

and the recipient-buyer in acute need might agree to pay any amount to save his life.

The clause which provides exceptions to the general rule against donation by non-relatives is too vast and vague to cover any situation. Likewise, the introduction of brain-stem death is bound to give rise to problems and lurking suspicion that it will lead to premature removal of organs. Especially in a country where the level of literacy and awareness is not high, the idea may not be well received. The need of the hour is to create an awareness to make bequests of organs. As observed by one doctor, "What does one do when there are no organs forthcoming from cadavers? Do I turn my patient away and say — prepare for your death?" With the augmentation of cadaver organs the need for live donors and trading would come down. The media can play an effective role; requests can be put in application forms for driving licenses, passports, insurance policies and other documents seeking consent. The organs of the dead are of no use to him anymore; it would be an act of highest magnanimity to donate them to give life to another person than dispose them of like carrion. **The donor is dead; long live the donor who passes over the torch of life to another.**

APPENDICES

APPENDIX I

The Delhi Artificial Insemination (Human) Act, 1995*
(Delhi Act No. 12 of 1995)
As passed by the Legislative Assembly of the National Capital Territory of Delhi

An Act to provide for the regulation of donation, sale and supply of human semen and ovum for the purpose of artificial insemination and for matters connected therewith or incidental thereto.

Be it enacted by the Legislative Assembly of the National Capital Territory of Delhi in the Forty-sixth year of the Republic of India as follows:-

1. (1) This Act may be called the Delhi Artificial Insemination (Human) Act, 1995.

(2) It extends to the whole of the National Capital Territory of Delhi.

(3) It shall come into force on such date as the Government may, by notification in the official Gazette, appoint.

2. In this Act, unless the context otherwise requires:

(a) "Delhi" means the National Capital Territory of Delhi;
(b) "donor" means the donor of semen, in the case of a male and of ovum, in the case of a female;
(c) "Government" means the Government of the National Capital Territory of Delhi;
(d) "Government Hospital" means any hospital established or maintained by the Government, or the Municipal Corporation of Delhi, or the New Delhi Municipal Council, performing the function of artificial insemination and includes any other hospital which may be declared by the Government, by notification in the official Gazette, to be a Government hospital for the purposes of this Act;
(e) "Hospital" means any premises including a maternity home, nursing home, hospital or any other place not established or maintained by Government funds, used or intended to be used for storage, supply or artificial insemination of semen;
(f) "Qualified Medical Practitioner" means a medical practitioner registered in any State or Union Territory in India under a law for the registration of medical practitioners;
(g) "Registration" means the registration under section 3 of this Act and the expression "registered" shall be construed accordingly;

* Assented by the President of India on 30th November 1996 and published in the Delhi Gazette, Extraordinary Part IV.

(h) "Semen" wherever it occurs in this Act, means the semen or ovum of male & female human being, as the case may be;

(i) "Semen Bank" means any premises used or intended to be used for storage, sale, donation, or supply of semen;

(j) "Supervisory Authority" means the Director of Health Services, Government of National Capital Territory of Delhi.

3. No person shall carry on a semen bank in Delhi unless he has been duly registered in respect of such semen bank and the registration in respect thereof has not been cancelled under section 7 of this Act.

4. (1) Every person intending to carry on a semen bank in Delhi shall make, every year, an application for registration or the renewal thereof, to the Supervisory Authority:

Provided that in case of semen bank which is in existence at the date of the commencement of this Act, an application for registration under this Act, shall be made within three months from the date of such commencement.

(2) Every application for registration or renewal of registration shall be made on such date and in such form and shall be accompanied by such fee, as may be prescribed.

5. (1) Subject to the provisions of this Act and the rules made thereunder, the Supervisory Authority shall, on receipt of an application for registration, register the applicant in respect of the semen bank named in the application and issue to him a certificate of registration in the prescribed form:

Provided that the Supervisory Authority may refuse to register the applicant if it is satisfied —

(a) that the applicant or any person employed by him at the semen bank, is not a fit person to carry on or to be employed at the semen bank named in the application; or

(b) that it does not have the qualified staff or equipment to carry out the prescribed tests of the donor/recipient or; to store the semen; or

(c) that for reasons connected with the situation, the construction, staff or equipment of the semen bank or any premises used in connection therewith, is or are not fit to be used for semen bank or such description as the semen bank mentioned in application or that the semen bank or premises is or are used or to be used for purposes which are, in any way, improper, or undesirable in the case of such semen bank.

6. Whoever contravenes the provisions of section 3 shall be punishable with fine, which may extend to five thousand rupees and in the case of second or subsequent offence, with imprisonment for a term which may extend to three months, or with fine which may extend to five thousand rupees.

7. Subject to the provisions of this Act, the Supervisory Authority may, at any time, cancel the registration of a person in respect of any semen bank

on any ground which would entitle it to refuse an application for the registration of that person in respect of that semen bank or on the ground that the person has been convicted of an offence under this Act or that any other person has been convicted of such an offence in respect of that semen bank.

8. (1) Before making an order refusing an application for registration or an order cancelling any registration, the Supervisory Authority shall give to the applicant or to the person registered as the case may be, not less than one month's notice of its intention to make such an order, and every such notice shall state the grounds on which the Supervisory Authority intends to make the order and shall, before making the order, give him (in person or by a representative) an opportunity of showing cause why the order should not be made.

(2) If the Supervisory Authority, after giving the applicant or the person registered, an opportunity of showing cause as aforesaid, decides to refuse the application for registration or to cancel the registration, as the case may be, it shall make an order to that effect and shall send a copy of the order by registered post to the applicant or the person registered.

(3) Any person aggrieved by an order refusing an application for registration or cancelling any registration may, within a month after the date on which the copy of the order was sent to him, appeal to the Government against such order or refusal. The decision of the Government on any such appeal shall be final.

(4) No such order shall come into force until after the expiration of a month from the date on which it was made or, where notice of appeal is given .against it, until the appeal has been decided or withdrawn.

9. (1) The Supervisory Authority or any officer empowered by it in this behalf may, subject to such general or special orders as may be made by the Government, enter and inspect any premises which are used or which the Supervisory Authority or the officer empowered by it has reasonable cause to believe that such premises may be used for the purpose of semen bank and may also enter Government hospital or any hospital and inspect any records required to be kept thereat in accordance with the provisions of this Act.

(2) If any pers n refuses to allow the Supervisory Authority or the officer empowered by it to enter or inspect any such premises or Government hospital or hospital as aforesaid, or to inspect any such records as aforesaid or obstructs the Supervisory Authority or the officer empowered by it in the execution of the powers under this section, he shall be guilty of an offence under this Act.

10. (1) The semen bank before accepting the semen for artificial insemination shall test the donor for presence of HIV 1 and 2 antibodies by using a highly sensitive ELISA Kit and if found negative, only then, the donor shall be allowed to donate.

Explanation

The expression "HIV 1 and 2, ELISA Kit or ELISA Test" and "HIV" used in this section or in other sections of this Act denote respectively

"Human Immune Deficiency Virus' Type 1 and Type 2", "Enzyme Linked Immune-sorbent Kit or Enzyme Linked Immune-sorbent Assay Test" and "Human Immune Deficiency Virus".

(2) The donor shall be screened for HIV surface antigen and if found negative, only then, the donor shall be allowed to donate.

(3) The semen bank shall keep complete bio-data, including mark of identification, of the donor.

11. The donated semen shall be stored either by cryo preservation of liquid nitrogen freezing or any other safe method for a period of minimum three months in order to exclude window period of HIV 1 and 2 infection in the donor.

12. At the end of three months, a second ELISA Test shall, by the same method, be performed on the donor.

13. If, after conducting of necessary test, the donor is found fit, only then, the semen shall be used.

14. The qualified medical practitioner or Government hospital or hospital or the semen bank performing artificial insemination, as the case may be, shall—

(a) keep complete record of the bio-data including mark of identification of the donor and the recipient of the semen or ovum;
(b) test the recipient for "HIV 1 and 2" and sexually transmitted diseases before performing artificial insemination;
(c) seek the written consent of the husband and the wife, seeking artificial insemination;
(d) seek the written consent of the donor and the recipient and their spouse, in case of requests of semen or ovum from specified donor/recipient;
(e) not segregate the XX or XY chromosomes for artificial insemination;

Explanation
The letters "XX" and "XY" used in this clause denote "female sex" and "male sex" respectively;

(f) maintain secrecy about the identity of the donor and the recipient of the semen/ovum;
(g) seek the written consent of the recipient for using the semen on the basis of only one ELISA Test being negative where facilities for cyro-preservation and liquid nitrogen for semen are not available.

15. If the Supervisory Authority has reason to believe that any of the provisions of this Act is being violated, he may call for any information or may seize any article, medicine or any other related things, admission register or other document maintained/kept or found at the place.

16. Whoever contravenes any provision of this Act or of any rule or order or requisition made thereunder shall, if no punishment is provided for the offence, be punishable for the first offence with fine which may extend to five hundred rupees, and for second or subsequent offence with fine which may extend to two thousand rupees.

17. Whoever contravenes any of the provisions of sections 10, 11, 12, 13 or 14 of this Act shall be punishable with imprisonment for a term which may extend to three years and with minimum fine of five thousand rupees.

18. No court other than the court of a Metropolitan Magistrate shall take cognizance of, and try an offence under this Act.

(2) No court shall take cognizance of any offence under this Act except on a complaint in writing of the Supervisory Authority or any Officer authorised by it, in writing, in this behalf.

(3) Notwithstanding anything contained in the Code of Criminal Procedure, 1973 (2 of 1974), offences under this Act shall be non-cognizable and bailable.

19. The Government may, by notification in the official Gazette, direct that any power exercisable by it under this Act, may also be exercised by such officer as may be mentioned therein, subject to such conditions, if any, as may be specified therein.

20. No suit, prosecution, or other legal proceedings shall lie against any officer, Supervisory Authority or Government for anything which is in good faith done or intended to be done in pursuance of this Act.

21. (1) The Government may, by notification in the Official Gazette, make rules to carry out the purposes of this Act.

(2) In particular, and without prejudice to the generality of the foregoing power, such rules may provide for all or any of the following matters, namely:

(a) prescription of date and form of application and the fee required to be paid for registration of semen banks;
(b) prescription of form of certificate of registration;
(c) prescription of tests in respect of donors/recipients;
(d) any other matter in respect of which this Act makes no provision or makes insufficient provision and provision is, in the opinion of the Government necessary.

(3) Any rule made by the Government shall be subject to previous publication thereof in the official Gazette.

(4) Every rule made under this Act shall be laid as soon as may be after it is made before the House of the Legislative Assembly and if the House agrees in making any modifications in the rule or the House agrees that the rule should not be made, the rule shall thereafter have effect only in such modified form or be of no effect, as the case may be, so, however, that any such modification or annulment shall be without prejudice to the validity of anything previously done under that rule.

APPENDIX II

The Pre-natal Diagnostic Techniques (Regulation and Prevention of Misuse) Act, 1994 No. 57 of 1994

[20th September, 1994]

An Act to provide for the regulation of the use of pre-natal diagnostic techniques for the purpose of detecting genetic or metabolic disorders or chromosomal abnormalities or certain congenital malformations or sex linked disorders and for the prevention of the misuse of such techniques for the purpose of pre-natal sex determination leading to female foeticide, and for matters connected there with or incidental thereto.

Be it enacted by Parliament in the Forty-fifth Year of the Republic of India as follows:

CHAPTER I

Preliminary

1. (1) This Act may be called the Pre-natal Diagnostic Techniques (Regulation and Prevention of Misuse) Act, 1994.

(2) It shall extend to the whole of India except the State of Jammu and Kashmir.

(3) It shall come into force on such date as the Central Government may, by notification in the Official Gazette, appoint.

2. In this Act, unless the context otherwise requires,—

(a) "Appropriate Authority" means the Appropriate Authority appointed under section 17;

(b) "Board" means the Central Supervisory Board constituted under section 7;

(c) "Genetic Counselling Centre" means an institute, hospital, nursing home or any place, by whatever name called, which provides for genetic counselling to patients;

(d) "Genetic Clinic" means a clinic, institute, hospital, nursing home or any place, by whatever name called, which is used for conducting pre-natal diagnostic procedures;

(e) "Genetic Laboratory" means a laboratory and includes a place where facilities are provided for conducting analysis or tests of samples received from Genetic Clinic for pre-natal diagnostic test;

(f) "gynaecologist" means a person who possesses a post-graduate qualification in gynaecology and obstetrics;

(g) "medical geneticist" means a person who possesses a degree or diploma or certificate in medical genetics in the field of pre-natal diagnostic techniques or has experience of not less than two years in such field after obtaining—

(i) any one of the medical qualifications recognised under the Indian Medical Council Act, 1956; or
(ii) a post-graduate degree in biological sciences;

(h) "paediatrician" means a person who possesses a post-graduate qualification in paediatrics;
(i) "pre-natal diagnostic procedures" means all gynaecological or obstetrical or medical procedures such as ultrasonography foetoscopy, taking or removing samples of amniotic fluid, chorionic villi, blood or any tissue of a pregnant woman for being sent to a Genetic Laboratory or Genetic Clinic for conducting pre-natal diagnostic test;
(j) "pre-natal diagnostic techniques" includes all pre-natal diagnostic procedures and pre-natal diagnostic tests;
(k) "pre-natal diagnostic test" means ultrasonography or any test or analysis of amniotic fluid, chorionic villi, blood or any tissue of a pregnant woman conducted to detect genetic or metabolic disorders or chromosomal abnormalities or congential anomalies or haemoglobinopathies or sex-linked diseases;
(l) "prescribed" means prescribed by rules made under this Act;
(m) "registered medical practitioner" means a medical practitioner who possesses any recognised medical qualification as defined in clause (h) of section 2 of the Indian Medical Council Act, 1956, and whose name has been entered in a State Medical Register;
(n) "regulations" means regulations framed by the Board under this Act.

CHAPTER II

Regulation of Genetic Counselling Centres Genetic Laboratories and Genetic Clinics

3. On and from the commencement of this Act —

(1) no Genetic Counselling Centre, Genetic Laboratory or Genetic Clinic unless registered under this Act, shall conduct or associate with, or help in, conducting activities relating to pre-natal diagnostic techniques;

(2) no Genetic Counselling Centre, Genetic Laboratory or Genetic Clinic shall employ or cause to be employed any person who does not possess the prescribed qualifications;

(3) no medical geneticist, gynaecologist paediatrician registered medical practitioner or any other person shall conduct or cause to be conducted or aid in conducting by himself or through any other person, any pre-natal diagnostic techniques at a place other than a place registered under this Act.

CHAPTER III

Regulation of Pre-natal Diagnostic Techniques

4. On and from the commencement of this Act—

(1) no place including a registered Genetic Counselling Centre or Genetic Laboratory or Genetic Clinic shall be used or caused to be used by any person for conducting pre-natal diagnostic techniques except for the purposes specified in clause (2) and after satisfying any of the conditions specified in clause (3);

(2) no pre-natal diagnostic techniques shall be conducted except for the purposes of detection of any of the following abnormalities, namely:

(i) chromosomal abnormalities;
(ii) genetic metabolic diseases;
(iii) haemoglobinopathies;
(iv) sex-linked genetic diseases;
(v) congenital anomalies;
(vi) any other abnormalities or diseases as may be specified by the Central Supervisory Board;

(3) no pre-natal diagnostic techniques shall be used or conducted unless the person qualified to do so is satisfied that any of the following conditions are fulfilled, namely:

(i) age of the pregnant woman is above thirty-five years;
(ii) the pregnant woman has undergone two or more spontaneous abortions or foetal loss;
(iii) the pregnant woman had been exposed to potentially teratogenic agents such as drugs, radiation, infection or chemicals;
(iv) the pregnant woman has a family history of mental retardation or physical deformities such as spasticity or any other genetic disease;
(v) any other condition as may be specified by the Central Supervisory Board;

(4) no person, being a relative or the husband of the pregnant woman shall seek or encourage the conduct of any pre-natal diagnostic techniques on her except for the purpose specified in clause (2).

5. (1) No person referred to in clause (2) of section 3 shall conduct the pre-natal diagnostic procedures unless —

(a) he has explained all known side and after effects of such procedures to the pregnant woman concerned;
(b) he has obtained in the prescribed form her written consent to undergo such procedures in the language which she understands; and
(c) a copy of her written consent obtained under clause (b) is given to the pregnant woman.

(2) No person conducting pre-natal diagnostic procedures shall communicate to the pregnant woman concerned or her relatives the sex of the foetus by words, signs or in any other manner.

6. On and from the commencement of this Act—

(a) no Genetic Counselling Centre or Genetic Laboratory or Genetic Clinic shall conduct or cause to be conducted in its Centre, Laboratory or Clinic, pre-natal diagnostic techniques including ultrasonography, for the purpose of determining the sex of a foetus;

(b) no person shall conduct or cause to be conducted any pre-natal diagnostic techniques including ultrasonography for the purpose of determining the sex of a foetus.

CHAPTER IV

Central Supervisory Board

7. (1) The Central Government shall constitute a Board to be known as the Central Supervisory Board to exercise the powers and perform the functions conferred on the Board under this Act.

(2) The Board shall consist of—

(a) the Minister in charge of the Ministry of Department of Family Welfare, who shall be the Chairman, *ex officio*;

(b) the Secretary to the Government of India in charge of the Department of Family Welfare, who shall be the Vice-Chairman, *ex-officio*;

(c) two members to be appointed by the Central Government to represent the Ministers of Central Government in charge of Woman and Child Development and of Law and Justice, *ex-officio*;

(d) the Director General of Health Services of the Central Government, *ex-officio*;

(e) ten members to be appointed by the Central Government, two each from amongst—

(i) eminent medical geneticists;
(ii) eminent gynaecologists and obstetricians;
(iii) eminent paediatricians;
(iv) eminent social scientists; and
(v) representatives of women welfare organisations;

(f) three women Members of Parliament, of whom two shall be elected by the House of the People and one by the Council of States;

(g) four members to be appointed by the Central Government by rotation to represent the States and the Union territories, two in the alphabetical order and two in the reverse alphabetical order:

Provided that no appointment under this clause shall be made except on the recommendation of the State Government or, as the case may be, the Union territory;

(h) an officer, not below the rank of a Joint Secretary or equivalent of the Central Government, in charge of Family Welfare, who shall be the Member-Secretary, *ex-officio*.

8. (1) The term of office of a member, other than an *ex-officio* member, shall be,—

(a) in case of appointment under clause (e) or clause (f) of sub-section (2) of section 7, three years; and
(b) in case of appointment under clause (g) of the said sub-section, one year.

(2) If a casual vacancy occurs in the office of any other members, whether by reason of his death, resignation or inability to discharge his functions owing to illness or other incapacity, such vacancy shall be filled by the Central Government by making a fresh appointment and the member so appointed shall hold office for the remainder of the term of office of the person in whose place he is so appointed.

(3) The Vice-Chairman shall perform such functions as may be assigned to him by the Chairman from time to time.

(4) The procedure to be followed by the members in the discharge of their functions shall be such as may be prescribed.

9. (1) The Board shall meet at such time and place, and shall observe such rules of procedure in regard to the transaction of business at its meetings (including the quorum at such meetings) as may be provided by regulations:

Provided that the Board shall meet at least once in six months.

(2) The Chairman and in his absence the Vice-Chairman shall preside at the meetings of the Board.

(3) If for any reason the Chairman or the Vice-Chairman is unable to attend any meeting of the Board, any other member chosen by the members present at the meeting shall preside at the meeting.

(4) All questions which come up before any meeting of the Board shall be decided by a majority of the votes of the members present and voting, and in the event of an equality of votes, the Chairman, or in his absence, the person presiding, shall have and exercise a second or casting vote.

(5) Members other than *ex-officio* members shall receive such allowances, if any, from the Board as may be prescribed.

10. No act or proceeding of the Board shall be invalid merely by reason of—

(a) any vacancy in, or any defect in the constitution of, the Board; or
(b) any defect in the appointment of a person acting as a member of the Board; or
(c) any irregularity in the procedure of the Board not affecting the merits of the case.

11. (1) The Board may associate with itself, in such manner and for such purposes as may be determined by regulations, any person whose assistance or advice it may desire in carrying out any of the provisions of this Act.

(2) A person associated with it by the Board under subsection (1) for any purpose shall have a right to take part in the discussions relevant to that

purpose, but shall not have a right to vote at a meeting of the Board and shall not be a member for any other purpose.

12. (1) For the purpose of enabling it efficiently to discharge its functions under this Act, the Board may, subject to such regulations as may be made in this behalf, appoint (whether on deputation or otherwise) such number of officers and other employees as it may consider necessary:

Provided that the appointment of such category of officers, as may be specified in such regulations, shall be subject to the approval of the Central Government.

(2) Every officer or other employee appointed by the Board shall be subject to such conditions of service and shall be entitled to such remuneration as may be specified in the regulations.

13. All orders and decisions of the Board shall be authenticated by the signature of the Chairman or any other member authorised by the Board in this behalf, and all other instruments issued by the Board shall be authenticated by the signature of the Member-Secretary or any other officer of the Board authorised in like manner in this behalf.

14. A person shall be disqualified for being appointed as a member if, he —

(a) has been convicted and sentenced to imprisonment for an offence which, in the opinion of the Central Government, involves moral turpitude; or
(b) is an undischarged insolvent; or
(c) is of unsound mind and stands so declared by a competent court; or
(d) has been removed or dismissed from the service of the Government or a Corporation owned or controlled by the Government; or
(e) has, in the opinion of the Central Government, such financial or other interest in the Board as is likely to affect prejudicially the discharge by him of his functions as a member; or
(f) has, in the opinion of the Central Government, been associated with the use or promotion of pre-natal diagnostic technique for determination of sex.

15. Subject to the other terms and conditions of service as may be prescribed, any person ceasing to be a member shall be eligible for reappointment as such member.

16. The Board shall have the following functions, namely:

(i) to advise the Government on policy matters relating to use of pre-natal diagnostic techniques;
(ii) to review implementation of the Act and the rules made thereunder and recommend changes in the said Act and rules to the Central Government;
(iii) to create public awareness against the practice of pre-natal determination of sex and female foeticide;
(iv) to lay down code of conduct to be observed by persons working at Genetic Counselling Centres, Genetic Laboratories and Genetic Clinics;
(v) any other functions as may be specified under the Act.

CHAPTER V

Appropriate Authority and Advisory Committee

17. (1) The Central Government shall appoint, by notification in the Official Gazette, one or more Appropriate Authorities for each of the Union territories for the purposes of this Act.

(2) The State Government shall appoint, by notification in the Official Gazette, one or more Appropriate Authorities for the whole or part of the State for the purposes of this Act having regard to the intensity of the problem of pre-natal sex determination leading to female foeticide.

(3) The officers appointed as Appropriate Authorities under sub-section (1) or sub-section (2) shall be,—

(a) when appointed for the whole of the State or the Union territory, of or above the rank of the Joint Director of Health and Family Welfare; and

(b) when appointed for any part of the State or the Union territory, of such other rank as the State Government or the Central Government, as the case may be, may deem fit.

(4) The Appropriate Authority shall have the following functions, namely:

(a) to grant, suspend or cancel registration of a Genetic Counselling Centre, Genetic Laboratory or Genetic Clinic;

(b) to enforce standards prescribed for the Genetic Counselling Centre, Genetic Laboratory and Genetic Clinic;

(c) to investigate complaints of breach of the provisions of this Act or the rules made thereunder and take immediate action; and

(d) to seek and consider the advice of the Advisory Committee, constituted under sub-section (5), on application for registration and on complaints for suspension or cancellation of registration.

(5) The Central Government or the State Government, as the case may be, shall constitute an Advisory Committee for each Appropriate Authority to aid and advise the Appropriate Authority in the discharge of its functions, and shall appoint one of the members of the Advisory Committee to be its Chairman.

(6) The Advisory Committee shall consist of—

(a) three medical experts from amongst gynaecologists, obstetricians, paediatricians and medical geneticists;

(b) one legal expert;

(c) one officer to represent the department dealing with information and publicity of the State Government or the Union territory, as the case may be;

(d) three eminent social workers of whom not less than one shall be from amongst representatives of women's organisations.

(7) No person who, in the opinion of the Central Government or the State Government, as the case may be, has been associated with the use or promotion of pre-natal diagnostic technique for determination of sex shall be appointed as a member of the Advisory Committee.

(8) The Advisory Committee may meet as and when it thinks fit or on the request of the Appropriate Authority for consideration of any application for registration or any complaint for suspension or cancellation of registration and to give advice thereon:

Provided that the period intervening between any two meetings shall not exceed the prescribed period.

(9) The terms and conditions subject to which a person may be appointed to the Advisory Committee and the procedure to be followed by such Committee in the discharge of its functions shall be such as may be prescribed.

CHAPTER VI

Registration of Genetic Counselling Centres, Genetic Laboratories and Genetic Clinics

18. (1) No person shall open any Genetic Counselling Centre, Genetic Laboratory or Genetic Clinic after the commencement of this Act unless such Centre, Laboratory or Clinic is duly registered separately or jointly under this Act.

(2) Every application for registration under sub-section (1), shall be made to the Appropriate Authority in such form and in such manner and shall be accompanied by such fees as may be prescribed.

(3) Every Genetic Counselling Centre, Genetic Laboratory or Genetic Clinic engaged, either partly or exclusively, in counselling or conducting pre-natal diagnostic techniques for any of the purposes mentioned in section 4, immediately before the commencement of this Act, shall apply for registration within sixty days from the date of such commencement.

(4) Subject to the provisions of section 6, every Genetic Counselling Centre, Genetic Laboratory or Genetic Clinic engaged in counselling or conducting pre-natal diagnostic techniques shall cease to conduct any such counselling or technique on the expiry of six months from the date of commencement of this Act unless such Centre, Laboratory or Clinic has applied for registration and is so registered separately or jointly or till such application is disposed or, whichever is earlier.

(5) No Genetic Counselling Centre, Genetic Laboratory or Genetic Clinic shall be registered under this Act unless the Appropriate Authority is satisfied that such Centre, Laboratory or Clinic is in a position to provide such facilities, maintain such equipment and standards as may be prescribed.

19. (1) The Appropriate Authority shall, after holding an inquiry and after satisfying itself that the applicant has complied with all the requirements of this Act and the rules made thereunder and having regard to the advice of the Advisory Committee in this behalf, grant a certificate of registration in the prescribed form jointly or separately to the Genetic Counselling Centre, Genetic Laboratory or Genetic Clinic, as the case may be.

(2) If, after the inquiry and after giving an opportunity of being heard to the applicant and having regard to the advice of the Advisory Committee, the Appropriate Authority is satisfied that the applicant has not complied with the requirements of this Act or the rules, it shall, for reasons to be recorded in writing, reject the application for registration.

(3) Every certificate of registration shall be renewed in such manner and after such period and on payment of such fees as may be prescribed.

(4) The certificate of registration shall be displayed by the registered Genetic Counselling Centre, Genetic Laboratory or Genetic Clinic in a conspicuous place at its place of business.

20. (1) The Appropriate Authority may *suo moto*, or on complaint, issue a notice to the Genetic Counselling Centre, Genetic Laboratory or Genetic Clinic to show cause why its registration should not be suspended or cancelled for the reasons mentioned in the notice.

(2) If, after giving a reasonable opportunity of being heard to the Genetic Counselling Centre, Genetic Laboratory or Genetic Clinic and having regard to the advice of the Advisory Committee, the Appropriate Authority is satisfied that there has been a breach of the provisions of this Act or the rules, it may, without prejudice to any criminal action that it may take against such Centre, Laboratory or Clinic, suspend its registration for such period as it may think fit or cancel its registration, as the case may be.

(3) Notwithstanding anything contained in sub-sections (1) and (2), if the Appropriate Authority is, of the opinion that it is necessary or expedient so to do in the public interest, it may, for reasons to be recorded in writing, suspend the registration of any Genetic Counselling Centre, Genetic Laboratory or Genetic Clinic without issuing any such notice referred to in sub-section (1).

21. The Genetic Counselling Centre, Genetic Laboratory or Genetic Clinic may, within thirty days from the date of receipt of the order of suspension or cancellation of registration passed by the Appropriate Authority under section 20, prefer an appeal against such order to —

(i) the Central Government, where the appeal is against the order of the Central Appropriate Authority; and

(ii) the State Government, where the appeal is against the order of the State Appropriate Authority, in the prescribed manner.

CHAPTER VII

Offences and Penalties

22. (1) No person, organisation, Genetic Counselling Centre, genetic Laboratory or Genetic Clinic shall issue or cause to be issued any advertisement in any manner regarding facilities of pre-natal determination of sex available at such Centre, Laboratory, Clinic or any other place.

(2) No person or organisation shall publish or distribute or cause to be published or distributed any advertisement in any manner regarding facilities of pre-natal determination of sex available at any Genetic Counselling Centre, Genetic Laboratory, Genetic Clinic or any other place.

(3) Any person who contravenes the provisions of sub-section (1) or sub-section (2) shall be punishable with imprisonment for a term which may extend to three years and with fine which may extend to ten thousand rupees.

Explanation,—For the purposes of this section, "advertisement" includes any notice, circular, label wrapper or other document and also includes any visible representation made by means of any light, sound, smoke or gas.

23. (1) Any medical geneticist, gynaecologist, registered medical practitioner or any person who owns a Genetic Counselling Centre, a Genetic Laboratory or a Genetic Clinic or is employed in such a Centre, Laboratory or Clinic and renders his professional or technical services to or at such a Centre, Laboratory or Clinic, whether on an honorary basis or otherwise, and who contravenes any of the provisions of this Act or rules made thereunder shall be punishable with imprisonment for a term which may extend to three years and with fine which may extend to ten thousand rupees and on any subsequent conviction, with imprisonment which may extend to five years and with fine which may extend to fifty thousand rupees.

(2) The name of the registered medical practitioner who has been convicted by the court under sub-section (1), shall be reported by the Appropriate Authority to the respective State Medical Council for taking necessary action including the removal of his name from the register of the Council for a period of two years for the first offence and permanently for the subsequent offence.

(3) Any person who seeks the aid of a Genetic Counselling Centre, Genetic Laboratory or Genetic Clinic or of a medical geneticist, gynaecologist or registered medical practitioner for conducting pre-natal diagnostic techniques on any pregnant woman (including such woman unless she was compelled to undergo such diagnostic techniques) for purposes other than those specified in clause (2) of section 4, shall, be punishable with imprisonment for a term which may extend to three years and with fine which may extend to ten thousand rupees and on any subsequent conviction with imprisonment which may extend to five years and with fine which may extend to fifty thousand rupees.

24. Notwithstanding anything in the Indian Evidence Act, 1872, the court shall presume unless the contrary is proved that the pregnant woman has been compelled by her husband or the relative to undergo pre-natal diagnostic technique and such person shall be liable for abetment of offence under sub-section (3) of section 23 and shall be punishable for the offence specified under that section.

25. Whoever contravenes any of the provisions of this Act or any rules made thereunder, for which no penalty has been elsewhere provided in this Act, shall be punishable with imprisonment for a term which may extend to three months or with fine, which may extend to one thousand rupees or with both and in the case of continuing contravention with an additional fine which may extend to five hundred rupees for every day during which such contravention continues after conviction for the first such contravention.

26. (1) Where any offence, punishable under this Act has been committed by a company, every person who, at the time the offence was committed was in charge of, and was responsible to, the company for the conduct of the business of the company, as well as the company, shall be deemed to be guilty of the offence and shall be liable to be proceeded against and punished accordingly:

Provided that nothing contained in this sub-section shall render any such person liable to any punishment, if he proves that the offence was committed without his knowledge or that he had exercised all due diligence to prevent the commission of such offence.

(2) Notwithstanding anything contained in sub-section (1), where any offence punishable under this Act has been committed by a company and it is proved that the offence has been committed with the consent or connivance of, or is attributable to any neglect on the part of, any director, manager, secretary or other officer of the company, such director, manager, secretary or other officer shall also be deemed to be guilty of that offence and shall be liable to be proceeded against and punished accordingly.

Explanation. — For the purposes of this section —

(a) "company" means any body corporate and includes a firm or other association of individuals, and

(b) "director", in relation to a firm, means a partner in the firm.

27. Every offence under this Act shall be cognizable, non-bailable and non-compoundable.

28. (1) No court shall take cognizance of an offence under this Act except on a complaint made by —

(a) the Appropriate Authority concerned, or any officer authorised in this behalf by the Central government or State Government, as the case may be, or the Appropriate Authority; or

(b) a person who has given notice of not less than thirty days in the manner prescribed to the Appropriate Authority, of the alleged offence and of his intention to make a complaint to the court.

Explanation — For the purpose of this clause, "person" includes a social organisation.

(2) No court other than that of a Metropolitan Magistrate or a Judicial Magistrate of the first class shall try any offence punishable under this Act.

(3) Where a complaint has been made under clause (b) of subsection (1), the court may, on demand by such person, direct the Appropriate Authority to make available copies of the relevant records in its possession to such person.

CHAPTER VIII

Miscellaneous

29. (1) All records, charts, forms, reports, consent letters and all other documents required to be maintained under this Act and the rules shall be preserved for a period of two years or for such period as may be prescribed:

Provided that, if any criminal or other proceedings are instituted against any Genetic Counselling Centre, Genetic Laboratory or Genetic Clinic, the records and all other documents of such Centre, Laboratory or Clinic shall be preserved till the final disposal of such proceedings.

(2) All such records shall, at all reasonable times, be made available for inspection to the Appropriate Authority or to any other person authorised by the Appropriate Authority in this behalf.

30. (1) If the Appropriate Authority has reason to believe that an offence under this Act has been or is being committed at any Genetic Counselling Centre, Genetic Laboratory or Genetic Clinic, such Authority or any officer authorised thereof in this behalf may, subject to such rules as may be prescribed, enter and search at all reasonable times with such assistance, if any, as such authority or officer considers necessary, such Genetic Counselling Centre, Genetic Laboratory or Genetic Clinic and examine any record, register, document, book, pamphlet, advertisement or any other material object found therein and seize the same if such Authority or officer has reason to believe that it may furnish evidence of the commission of an office punishable under this Act.

(2) The provisions of the Code of Criminal Procedure, 1973 relating to searches and seizures shall, so far as may be, apply to every search or seizure made under this Act.

31. No suit, prosecution or other legal proceeding shall lie against the Central or the State Government or the Appropriate Authority or any officer authorised by the Central or State Government or by the Authority for anything which is in good faith done or intended to be done in pursuance of the provisions of this Act.

32. (1) The Central Government may make rules for carrying out the provisions of this Act.

(2) In particular and without prejudice, to the generality of the foregoing power, such rules may provide for—

(i) the minimum qualifications for persons employed at a registered Genetic Counselling Centre, Genetic Laboratory or Genetic Clinic under clause (1) of section 3;

(ii) the form in which consent of a pregnant woman has to be obtained under section 5;

(iii) the procedure to be followed by the members of the Central Supervisory Board in the discharge of their functions under sub-section (4) of section 8;

(iv) allowances for members other than *ex-officio* members admissible under sub-section (5) of section 9;

(v) the period intervening between any two meetings of the Advisory Committee under the provision to sub-section (8) of section 17;

(vi) the terms and conditions subject to which a person may be appointed by the Advisory Committee and the procedure to be followed by such Committee under sub-section (9) of section 17;

(vii) the form and manner in which an application shall be made for registration and the fee payable thereof under sub-section (2) of section 18;

(viii) the facilities to be provided, equipment and other standards to be maintained by the Genetic Counselling Centre, Genetic Laboratory or Genetic Clinic under sub-section (5) of section 18;

(ix) the form in which a certificate of registration shall be issued under sub-section (1) of section 19;

(x) the manner in which and the period after which a certificate of registration shall be renewed and the fee payable for such renewal under sub-section (3) of section 19;

(xi) the manner in which an appeal may be preferred under section 21;

(xii) the period up to which records, charts, etc., shall be preserved under sub-section (1) of section 29;

(xiii) the manner in which the seizure of documents, records, objects, etc., shall be made and the manner in which seizure list shall be prepared and delivered to the person from whose custody such documents, records or objects were seized under sub-section (1) of section 30;

(xiv) any other matter that is required to be, or may be prescribed.

33. The Board may, with the previous sanction of the Central Government, by notification in the Official Gazette, make regulations not inconsistent with the provisions of this Act and the rules made thereunder to provide for —

(a) the time and place of the meetings of the Board and the procedure to be followed for the transaction of business at such meetings and the number of members which shall form the quorum under sub-section (1) of section 9;

(b) the manner in which a person may be temporarily associated with the Board under sub-section (1) of section 11;

(c) the method of appointment, the conditions of service and the scales of pay and allowances of the officer and other employees of the Board appointed under section 12;

(d) generally for the efficient conduct of the affairs of the Board.

34. Every rule and every regulation made under this Act shall be laid, as soon as may be after it is made, before each House of Parliament, while it is in session, for a total period of thirty days which may be comprised in one session or in two or more successive sessions, and if, before the expiry of the session immediately following the session or the successive sessions aforesaid, both Houses agree in making any modification in the rule or regulation or both Houses agree that the rule or regulation should not be made, the rule or regulation shall thereafter have effect only in such modified form or be of no effect, as the case may be; so, however, that any such modification or annulment shall be without prejudice to the validity of anything previously done under that rule or regulation.

RULES

Ministry of Health and Family Welfare

No. G.S.R. 1 In exercise of the powers conferred by Section 32 of the Pre-natal Diagnostic Techniques (Regulation and Prevention of Mususe) Act,

1994 the Central Government hereby makes the following rules, namely:

1. *Short title and commencement.* (a) These rules may be called the Prenatal Diagnostic Techniques (Regulation and Prevention of Misuse) Rules, 1996.

(2) They shall come into force on the date of their publication in the Official Gazette.

2. *Definitions.* In these rules, unless the context otherwise requires,

(a) "Act" means the Prenatal Diagnostic Techniques (Regulation and Prevention of Misuse) Act, 1994 (57 of 1994);
(b) "employee" means a person working in or employed by a Genetic Counselling Center, a Genetic Laboratory or a Genetic Clinic, and includes those working part-time, contractual, consultancy, honorary or on any other basis;
(c) "Form" means a Form appended to these rules;
(d) "Schedule" means a Schedule appended to these rules;
(e) "section" means a section of the Act;
(f) words and expressions used herein and not defined in these rules but defined in the Act, shall have the meanings, respectively, assigned to them in the Act.

3. *Minimum requirements.* (1) The minimum qualifications of the employees, the minimum equipment and minimum place for a Genetic Counselling Centre, Genetic Laboratory and Genetic Clinic shall be as specified in Schedules I, II and III.

(2) Where an institute, hospital, nursing home, or any place, by whatever name called provides services jointly of Genetic Counselling Centre, Genetic Laboratory and Genetic Clinic, or any combination of these, it shall conform to the requirements as specified in Schedules I, II and III.

4. *Registration of Genetic Counselling Centre, Genetic Laboratory and Genetic Clinic.* (1) An application for registration shall be made to the Appropriate Authority, in duplicate, in Form A.

(2) The Appropriate Authority, or any person in his office authorised in this behalf, shall acknowledge receipt of the application for registration, in the acknowledgment slip provided at the bottom of Form A, immediately if delivered at the office of the Appropriate Authority, or not later than the next working day if received by post.

5. *Application Fee.* (1) Every application for registration under Rule 4 shall be accompanied by an application fee of:-

(a) Rs. 2000.00 for Genetic Counselling Centre;
(b) Rs. 3000.00 for Genetic Laboratory;
(c) Rs. 3000.00 for Genetic Clinic; and
(d) Rs. 4000.00 for an institute, hospital, nursing home or any place providing jointly the services of a Genetic Counselling Centre, Genetic Laboratory and Genetic Clinic or any combination of such Centre, Laboratory or Clinic.

(2) The application fee shall be paid by a demand draft drawn in favour of the Appropriate Authority, on any scheduled bank located at the headquarters of the Appropriate Authority.

6. *Certificate of registration.* (1) The Appropriate Authority shall, after making such enquiry and after satisfying itself that the applicant has complied with all the requirements, place the application before the Advisory Committee for its advice.

(2) Having regard to the advice of the Advisory Committee the Appropriate Authority shall grant a certificate of registration, in duplicate, in Form B to the applicant. One copy of the certificate of registration shall be displayed by the registered Genetic Counselling Centre, Genetic Laboratory or Genetic Clinic at a conspicuous place at its place of business:

Provided that the Appropriate Authority may grant a certificate of registration to a Genetic Laboratory or a Genetic Clinic to conduct one or more specified prenatal diagnostic tests or procedures, depending on the availability of place, equipment and qualified employees, and standards maintained by such laboratory or clinic.

(3) If, after enquiry and after giving an opportunity of being heard to the applicant and having regard to the advice of the Advisory Committee, the Appropriate Authority is satisfied that the applicant has not complied with the requirements of the Act and these rules, it shall, for reasons to be recorded in writing, reject the application for registration and communicate such rejection to the applicant specified in Form C.

(4) An enquiry under sub-rule (1), including inspection at the premises of the Genetic Counselling Centre, Genetic Laboratory or Genetic Clinic, shall be carried out only after due notice is given to the applicant by the Appropriate Authority.

(5) Grant of certificate of registration or rejection of application for registration shall be communicated to the applicant as specified in Form B or Form C, as the case may be, within a period of ninety days from the date of receipt of application for registration.

(6) The certificate of registration shall be non-transferable. In the event of change of ownership or change of management or on ceasing to function as Genetic Counselling Centre, Genetic Laboratory or Genetic Clinic, both copies of the certificate of registration shall be surrendered to the Appropriate Authority.

(7) In the event of change of ownership or change of management of Genetic Counselling Centre, Genetic Laboratory or Genetic Clinic, the new owner or manager of such Centre, Laboratory or Clinic shall apply afresh for grant of certificate of registration.

7. *Validity of registration.* Every certificate of registration shall be valid for a period of five years from the date of its issue.

8. *Renewal of registration.* (1) An application for renewal of certificate of registration shall be made in duplicate in Form A, to the Appropriate Authority thirty days before the date of expiry of the certificate of registration. Acknowledgment of receipt of such application shall be issued by the Appropriate Authority in the manner specified in sub-rule (2) of Rule 4.

(2) The Appropriate Authority shall, after holding an enquiry and after satisfying itself that the applicant has complied with all the requirements of the Act and these rules and having regard to the advice of the Advisory Committee in this behalf renew the certificate of registration, as specified in

Form B, for a further period of five years from the date of expiry of the certificate of registration earlier granted.

(3) If, after enquiry and after giving an opportunity of being heard to the applicant and having regard to the advice of the Advisory Committee, the Appropriate Authority is satisfied that the applicant has not complied with the requirement of the Act and these rules, it shall, for reasons to be recorded in writing, reject the application for renewal of certificate of registration and communicate such rejection to the applicant as specified in Form C.

(4) The fees payable for renewal of certificate of registration shall be one of the fees provided in sub-rule (1) of Rule 5.

(5) On receipt of the renewed certificate of registration in duplicate or on receipt of communication or rejection of application for renewal, both copies of earlier certificate of registration shall be surrendered immediately to the Appropriate Authority by the Genetic Counselling Centre, Genetic Laboratory or Genetic Clinic.

(6) In the event of failure of the Appropriate Authority to renew the certificate of registration or to communicate rejection of application for renewal of registration within a period of ninety days from the date of receipt of application for renewal of registration, the certificate of registration shall be deemed to have been renewed.

9. *Maintenance and preservation of records.* (1) Every Genetic Counselling Centre, Genetic Laboratory and Genetic Clinic shall maintain a register showing serial order, the names and addresses of the women given genetic counselling, subjected to prenatal diagnostic procedures or prenatal diagnostic tests, the names of their husbands or fathers and the date on which they first reported for such counselling, procedure or test.

(2) The record to be maintained by every Genetic Counselling Centre, in respect of each woman counselled, shall be as specified in Form D.

(3) The record to be maintained by every Genetic Laboratory, in respect of each woman subjected to any prenatal diagnostic test, shall be as specified in Form E.

(4) The record to be maintained by every Genetic Clinic, in respect of each woman subjected to any prenatal diagnostic procedure, shall be as specified in Form F.

(5) The Appropriate Authority shall maintain a permanent record of applications for grant or renewal of certificate of registration as specified in Form H. Letters of intimation of every change of employee, place, address and equipment installed shall also be preserved as permanent records.

(6) All case-related records, forms of consent, laboratory results, microscopic pictures, sonographic plates or slides, recommendations and letters shall be preserved by the Genetic Counselling Centre, Genetic Laboratory or Genetic Clinic for a period of two years from the date of completion of counselling, pre-natal diagnostic procedure or prenatal diagnostic test, as the case may be. In the event of any legal proceedings, the records shall be preserved till the final disposal of legal proceedings, or till the expiry of the said period of two years, whichever is later.

(7) In case the Genetic Counselling Centre or Genetic Laboratory or Genetic Clinic maintains records on computer or other electronic equipment a printed copy of the record shall be taken and preserved after authentication by a person responsible for such record.

10. *Conditions for conducting prenatal diagnostic procedures.* (1) Before conducting any prenatal diagnostic procedure, a written consent, as specified in Form G, in a language the pregnant woman understands, shall be obtained from her:

Provided that where a Genetic Clinic has taken a sample of any body tissue or body fluid and sent it to a Genetic Laboratory for analysis or test, it shall not be necessary for the Genetic Laboratory to obtain a fresh consent in Form G.

(2) All the State Governments and Union territories may issue translation of Form G in languages used in the State or Union territory and where no official translation in a language understood by the pregnant woman is available, the Genetic Clinic may translate Form G into a language she understands.

11. *Facilities for inspection.* Every Genetic Counselling Centre, Genetic Laboratory and Genetic Clinic shall afford reasonable facilities for inspection of the place, equipment and records to the Appropriate Authority or to any other person authorised by the Appropriate Authority in this behalf.

12. *Procedure for search and seizure.* (1) The Appropriate Authority or any officer authorised in this behalf may enter and search at all reasonable times any Genetic Counselling Centre, Genetic Laboratory or Genetic Clinic, in the presence of two or more independent and respectable persons, for the purposes of Section 30.

(2) A list of any document, record, register, book, pamphlet, advertisement or any other material object found in the Genetic Counselling Centre, Genetic Laboratory or Genetic Clinic and seized shall be prepared in duplicate at the place of effecting the seizure. Both copies of such list shall be signed on every page by the Appropriate Authority or the officer authorised in this behalf and by the witnesses to the seizure:

Provided that the list may be prepared, in the presence of the witnesses, at a place other than the place of seizure if, for reasons to be recorded in writing, it is not practicable to make the list at the place of effecting the seizure.

(3) One copy of the list referred to in sub-rule (2) shall be handed over, under acknowledgement, to the person from whose custody the document, record, register, book, pamphlet, advertisement or any other material object have been seized:

Provided that a copy of the list of such document, record, register, book, pamphlet, advertisement or other material object seized may be delivered under acknowledgement, or sent by registered post to the owner or manager of the Genetic Counselling Centre, Genetic Laboratory or Genetic Clinic, if no person acknowledging custody of the document, record, register, book, pamphlet, advertisement or other material object seized is available at the place of effecting the seizure.

(4) If any material object seized is perishable in nature, the Appropriate Authority, or the officer authorised in this behalf shall make arrangements promptly for sealing, identification and preservation of the material object and also convey it to a facility for analysis or test, if analysis or test be required:

Provided that the refrigerator or other equipment used by the Genetic Counselling Centre, Genetic Laboratory or Genetic Clinic for preserving such perishable material object may be sealed until such time as arrangements can be made for safe removal of such perishable material object and in such eventuality, mention of keeping the material object seized, on the premises of the Genetic Counselling Centre, Genetic Laboratory or Genetic Clinic shall be made in the list of seizure.

(5) In the case of non-completion of search and seizure operation, the Appropriate Authority or the officer authorized in this behalf may make arrangement by way of mounting a guard or sealing of the premises of the Genetic Counselling Centre, Genetic Laboratory or Genetic Clinic, for safe keeping, listing and removing of documents, records, books or any other material object to be seized, and prevent any tampering with such documents, records, books or any other material object.

13. *Intimation of changes in employees, place or equipment.* Every Genetic Counselling Centre, Genetic Laboratory or Genetic Clinic shall intimate every change of employee, place, address and equipment installed, to the Appropriate Authority within a period of thirty days of such change.

14. *Conditions for analysis or test and prenatal diagnostic procedures.* (1) No Genetic Laboratory shall accept for analysis or test any sample, unless referred by a Genetic Clinic.

(2) Every prenatal diagnostic procedure shall invariably be immediately preceded by locating the foetus and placenta through ultrasonography, and prenatal diagnostic procedure shall be done under direct ultrasonographic monitoring so as to prevent any damage to the foetus and placenta.

15. *Meetings of the Advisory Committees.* The intervening period between any two meetings of Advisory Committees constituted under sub-section (5) of section 17 to advise the Appropriate Authority shall not exceed sixty days.

16. *Allowances to members of the Central Supervisory Board.* (1) The ex officio members, and other Central and State Government officers appointed by the Board will be entitled to Travelling Allowance and Daily Allowance for attending the meetings of the Board as per the Travelling Allowances rules applicable.

(2) The non-official members appointed to, and Members of Parliament elected to, the Board will be entitled to Travelling Allowance and Daily Allowance for attending the meetings of the Board as admissible to non-officials and Members of Parliament, as the case may be, under the Travelling Allowances rules of the Central Government.

17. *Public Information.* (1) Every Genetic Counselling Centre, Genetic Laboratory and Genetic Clinic shall prominently display on its premises a notice in English and in the local language or languages for the

information of the public, to the effect that disclosure of the sex of the foetus is prohibited under law.

(2) At least one copy each of the Act and these rules shall be available on the premises of every Genetic Counselling Centre, Genetic Laboratory and Genetic Clinic and shall be made available to the clientele on demand for perusal.

(3) The Appropriate Authority, the Central Government, the State Government and the Government/Administration of the Union territory may publish periodically lists of registered Genetic Counselling Centres, Genetic Laboratories and Genetic Clinics and findings from the reports and other information in their possession, for the information of the public and for use by the experts in the field.

[Schedules and Forms Omitted]

The Transplantation of Human Organs Act, 1994
(No. 42 of 1994)
[8th July, 1994]

An Act to provide for the regulation of removal, storage and transplantation of human organs for therapeutic purposes and for the prevention of commercial dealings in human organs and for matters connected therewith or incidental thereto.

Whereas it is expedient to provide for the regulation of removal, storage and transplantation of human organs for therapeutic purposes and for the prevention of commercial dealings in human organs;

And whereas Parliament has no power to make laws for the States with respect to any of the matters aforesaid except as provided in articles 249 and 250 of the Constitution;

And whereas in pursuance of clause (1) of article 252 of the Constitution, resolutions have been passed by all the Houses of the Legislatures of the States of Goa, Himachal Pradesh and Maharashtra to the effect that the matters aforesaid should be regulated in those States by Parliament by law;

Be it enacted by Parliament in the Forty-fifth Year of the Republic of India as follows:-

CHAPTER I

Preliminary

1. (1) This Act may be called the Transplantation of Human Organs Act, 1994.

(2) It applies, in the first instance, to the whole of the States of Goa, Himachal Pradesh and Maharashtra and to all the Union territories and it shall also apply to such other State which adopts this Act by resolution passed in that behalf under clause (1) of article 252 of the Constitution.

(3) It shall come into force in the States of Goa, Himachal Pradesh and Maharashtra and in all the Union territories on such date as the Central Government may, by notification, appoint and in any other State which adopts this Act under clause (1) of article 252 of the Constitution, on the date of such adoption; and any reference in this Act to the commencement of this Act shall, in relation to any State or Union territory, means the date on which this Act comes into force in such State or Union territory.

2. In this Act, unless the context otherwise requires —

(a) "advertisement" includes any form of advertising whether to the public generally or to any section of the public or individually to selected persons;
(b) "Appropriate Authority" means the Appropriate Authority appointed under section 13;
(c) "Authorisation Committee" means the committee constituted under clause (a) or clause (b) of sub-section (4) of section 9;
(d) "brain-stem death" means the stage at which all functions of the brain-stem have permanently and irreversibly ceased and is so certified under sub-section (6) of section 3;
(e) "deceased person" means a person in whom permanent disappearance of all evidence of life occurs, by reason of brain-stem death or in a cardio-pulmonary sense, at any time after live birth has taken place;
(f) "donor" means any person, not less than eighteen years of age, who voluntarily authorises the removal of any of his human organs for therapeutic purposes under sub-section (1) or sub-section (2) of section 3;
(g) "hospital" includes a nursing home, clinic, medical centre, medical or teaching institution for therapeutic purposes and other like institution;
(h) "human organ" means any part of a human body consisting of a structured arrangement of tissues which, if wholly removed, cannot be replicated by the body;
(i) "near relative" means spouse, son, daughter, father, mother, brother or sister;
(j) "notification" means a notification published in the Official Gazette;
(k) "payment" means payment in money or money's worth but does not include any payment for defraying or reimbursing —
 (i) the cost of removing, transporting or preserving the human organ to be supplied; or
 (ii) any expenses or loss of earnings incurred by a person so far as reasonably and directly attributable to his supplying any human organ from his body;
(l) "prescribed" means prescribed by rules made under this Act;
(m) "recipient" means a person into whom any human organ is, or is proposed to be, transplanted;
(n) "registered medical practitioner" means a medical practitioner who possesses any recognised medical qualification as defined in clause (h) of section 2 of the Indian Medical Council Act, 1956, and who is enrolled on a State Medical Register as defined in clause (k) of that section;
(o) "therapeutic purposes" means systematic treatment of any disease or the measures to improve health according to any particular method or modality; and
(p) "transplantation" means the grafting of any human organ from any living person or deceased person to some other living person for therapeutic purposes.

CHAPTER II

Authority for the Removal of Human Organs

3. (1) Any donor may, in such manner and subject to such conditions as may be prescribed, authorise the removal, before his death, of any human organ of his body for therapeutic purposes.

(2) If any donor had, in writing and in the presence of two or more witnesses (at least one of whom is a near relative of such person), unequivocally authorised at any time before his death, the removal of any human organ of his body, after his death, for therapeutic purposes, the person lawfully in possession of the dead body of the donor shall, unless he has any reason to believe that the donor had subsequently revoked the authority aforesaid, grant to a registered medical practitioner all reasonable facilities for the removal, for therapeutic purposes, of that human organ from the dead body of the donor.

(3) Where no such authority as is referred to in sub-section (2), was made by any person before his death but no objection was also expressed by such person to any of his human organs being used after his death for therapeutic purposes, the person lawfully in possession of the dead body of such person may, unless he has reason to believe that any near relative of the deceased person has objection to any of the deceased person's human organs being used for therapeutic purposes, authorise the removal of any human organ of the deceased person for its use for therapeutic purposes.

(4) The authority given under sub-section (1) or sub-section (2) or, as the case may be, sub-section (3) shall be sufficient warrant for the removal, for therapeutic purposes, of the human organ; but no such removal shall be made by any person other than the registered medical practitioner.

(5) Where any human organ is to be removed from the body of a deceased person, the registered medical practitioner shall satisfy himself, before such removal, by a personal examination of the body from which any human organ is to be removed, that life is extinct in such body or where it appears to be a case of brain-stem death, that such death has been certified under sub-section (6).

(6) Where any human organ is to be removed from the body of a person in the event of his brain-stem death, no such removal shall be undertaken unless such death is certified, in such form and in such manner and on satisfaction of such conditions and requirements as may be prescribed, by a Board of medical experts consisting of the following namely:

(i) the registered medical practitioner in charge of the hospital in which brain-stem death has occurred;

(ii) an independent registered medical practitioner, being a specialist, to be nominated by the registered medical practitioner specified in clause (i), from the panel of names approved by the Appropriate Authority;

(iii) a neurologist or a neurosurgeon to be nominated by the registered medical practitioner specified in clause (i), from the panel of names approved by the Appropriate Authority, and

(iv) the registered medical practitioner treating the person whose brain-stem death has occurred.

(7) Notwithstanding anything contained in sub-section (3), where brain-stem death of any person, less than eighteen years of ago, occurs and is certified under sub-section (6), any of the parents of the deceased person may give authority, in such form and in such manner as may be prescribed, for the removal of any human organ from the body of the deceased person.

4. (1) No facilities shall be granted under sub-section (2) of section 3 and no authority shall be given under sub-section (3) of that section for the removal of any human organ from the body of a deceased person, if the person required to grant such facilities, or empowered to give such authority, has reason to believe that an inquest may be required to be held in relation to such body in pursuance of the provisions of any law for the time being in force.

(2) No authority for the removal of any human organ from the body of a deceased person shall be given by a person to whom such body has been entrusted solely for the purpose of interment, cremation or other disposal.

5. (1) In the case of a dead body lying in a hospital or prison and not claimed by any of the near relatives of the deceased person within forty-eight hours from the time of the death of the concerned person, the authority for the removal of any human organ from the dead body which so remains unclaimed may be given, in the prescribed form, by the person in charge, for the time being, of the management or control of the hospital or prison, or by an employee of such hospital or prison authorised in this behalf by the person in charge of the management or control thereof.

(2) No authority shall be given under sub-section (1) if the person empowered to give such authority has reason to believe that any near relative of the deceased person is likely to claim the dead body even though such near relative has not come forward to claim the body of the deceased person within the time specified in sub-section (1).

6. Where the body of a person has been sent for post-mortem examination —

(a) for medico-legal purposes by reason of the death of such person having been caused by accident or any other unnatural cause; or
(b) for pathological purposes,

the person competent under this Act to give authority for the removal of any human organ from such dead body may, if he has reason to believe that such human organ will not be required for the purpose for which such body has been sent for postmortem examination, authorise the removal, for therapeutic purposes, of that human organ of the deceased person provided that he is satisfied that the deceased person had not expressed, before his death, any objection to any of his human organs being used, for therapeutic purposes after his death or, where he had granted an authority for the use of any of his human organs for therapeutic purposes after his death, such authority had not been revoked by him before his death.

7. After the removal of any human organ from the body of any person, the registered medical practitioner shall take such steps for the preservation of the human organ so removed as may be prescribed.

8. (1) Nothing in the foregoing provisions of this Act shall be construed as rendering unlawful any dealing with the body or with any part of the body of a deceased person if such dealing would have been lawful if this Act had not been passed.

(2) Neither the grant of any facility or authority for the removal of any human organ from the body of a deceased person in accordance with the provisions of this Act nor the removal of any human organ from the body of a deceased person in pursuance of such authority shall be deemed to be an offence punishable under section 297 of the Indian Penal Code.

9. (1) Save as otherwise provided in sub-section (3), no human organ removed from the body of a donor before his death shall be transplanted into a recipient unless the donor is a near relative of the recipient.

(2) Where any donor authorises the removal of any of his human organs after his death under sub-section (2) of section 3 or any person competent or empowered to give authority for the removal of any human organ from the body of any deceased person authorises such removal, the human organ may be removed and transplanted into the body of any recipient who may be in need of such human organ.

(3) If any donor authorises the removal of any of his human organs before his death under sub-section (1) of section 3 for transplantation into the body of such recipient, not being a near relative, as is specified by the donor by reason of affection or attachment towards the recipient or for any other special reasons, such human organ shall not be removed and transplanted without the prior approval of the Authorisation Committee.

(4) (a) The Central Government shall constitute, by notification, one or more Authorisation Committees consisting of such members as may be nominated by the Central Government on such terms and conditions as may be specified in the notification for each of the Union territories for the purposes of this section.

(b) The State Government shall constitute, by notification, one or more Authorisation Committees consisting of such members as may be nominated by the State Government on such terms and conditions as may be specified in the notification for the purposes of this section.

(5) On an application jointly made, in such form and in such manner as may be prescribed, by the donor and the recipient, the Authorisation Committee shall, after holding an inquiry and after satisfying itself that the applicants have complied with all the requirements of this Act and the rules made thereunder, grant to the applicants approval for the removal and transplantation of the human organ.

(6) If, after the inquiry and after giving an opportunity to the applicants of being heard, the Authorisation Committee is satisfied that the applicants have not complied with the requirements of this Act and the rules made thereunder, it shall, for reasons to be recorded in writing, reject the application for approval.

CHAPTER III

Regulation of Hospitals

10. (1) On and from the commencement of this Act —

(a) no hospital, unless registered under this Act, shall conduct, or associate with, or help in the removal, storage or transplantation of any human organ;
(b) no medical practitioner or any other person shall conduct, or cause to be conducted, or aid in conducting by himself or through any other person, any activity relating to the removal, storage or transplantation of any human organ at a place other than a place registered under this Act; and
(c) no place including a hospital registered under sub-section (1) of section 15 shall be used or cause to be used by any person for the removal, storage or transplantation of any human organ except for therapeutic purposes.

(2) Notwithstanding anything contained in sub-section (1), the eyes or the ears may be removed at any place from the dead body of any donor, for therapeutic purposes, by a registered medical practitioner.

Explanation.—For the purposes of this sub-section, "ears" includes ear drums and ear bones.

11. No donor and no person empowered to give authority for the removal of any human organ shall authorise the removal of any human organ for any purpose other than therapeutic purposes.

12. No registered medical practitioner shall undertake the removal or transplantation of any human organ unless he had explained, in such manner as may be prescribed, all possible effects, complications and hazards connected with the removal and transplantation to the donor and the recipient respectively.

CHAPTER IV

Appropriate Authority

13. (1) The Central Government shall appoint, by notification, one or more officers as Appropriate Authorities for each of the Union territories for the purposes of this Act.

(2) The State Government shall appoint, by notification, one or more officers as Appropriate Authorities for the purposes of this Act.

(3) The Appropriate Authority shall perform the following functions, namely—

(i) to grant registration under sub-section (1) of section 15 or renew registration under sub-section (3) of that section;
(ii) to suspend or cancel registration under sub-section (2) of section 16;

(iii) to enforce such standards, as may be prescribed, for hospitals engaged in the removal, storage or transplantation of any human organ;

(iv) to investigate any complaint of breach of any of the provisions of this Act or any of the rules made thereunder and take appropriate action;

(v) to inspect hospitals periodically for examination of the quality of transplantation and the follow-up medical care to persons who have undergone transplantation and persons from whom organs are removed; and

(vi) to undertake such other measures as may be prescribed.

CHAPTER V

Registration of Hospitals

14. (1) No hospital shall commence any activity relating to the removal, storage or transplantation of any human organ for therapeutic purposes after the commencement of this Act unless such hospital is duly registered under this Act:

Provided that every hospital engaged, either partly or exclusively, in any activity relating to the removal, storage or transplantation of any human organ for therapeutic purposes immediately before the commencement of this Act, shall apply for registration within sixty days from the date of such commencement:

Provided further that every hospital engaged in any activity relating to the removal, storage or transplantation of any human organ shall cease to engage in such activity on the expiry of three months from the date of commencement of this Act unless such hospital has applied for registration and is so registered or till such application is disposed of, whichever is earlier.

(2) Every application for registration under sub-section (1) shall be made to the Appropriate Authority in such form and in such manner and shall be accompanied by such fees as may be prescribed.

(3) No hospital shall be registered under this Act unless the Appropriate Authority is satisfied that such hospital is in a position to provide such specialised services and facilities, possess such skilled manpower and equipments and maintain such standards as may be prescribed.

15. (1) The Appropriate Authority shall, after holding an inquiry and after satisfying itself that the applicant has complied with all the requirements of this Act and the rules made thereunder, grant to the hospital a certificate of registration in such form, for such period and subject to such conditions as may be prescribed.

(2) If, after the inquiry and after giving an opportunity to the applicant of being heard, the Appropriate Authority is satisfied that the applicant has not complied with the requirements of this Act and the rules made thereunder, it shall, for reasons to be recorded in writing, reject the application for registration.

(3) Every certificate of registration shall be renewed in such manner and on payment of such fees as may be prescribed.

16. (1) The Appropriate Authority may, *suo moto* or on complaint, issue a notice to any hospital to show cause why its registration under this Act should not be suspended or cancelled for the reasons mentioned in the notice.

(2) If, after giving a reasonable opportunity of being heard to the hospital, the Appropriate Authority is satisfied that there has been a breach of any of the provisions of this Act or the rules made thereunder, it may, without prejudice to any criminal action that it may take against such hospital, suspend its registration for such period as it may think fit or cancel its registration:

Provided that where the Appropriate Authority is of the opinion that it is necessary or expedient so to do in the public interest, it may, for reasons to be recorded in writing, suspend the registration of any hospital without issuing any notice.

17. Any person aggrieved by an order of the Authorisation Committee rejecting an application for approval under sub-section (6) of section 9, or any hospital aggrieved by an order of the Appropriate Authority rejecting an application for registration under sub-section (2) of section 15 or an order of suspension or cancellation of registration under sub-section (2) of section 16, may, within thirty days from the date of the receipt of the order, prefer an appeal, in such manner as may be prescribed, against such order to—

(i) the Central Government where the appeal is against the order of the Authorisation Committee constituted under clause (a) of sub-section (4) of section 9 or against the order of the Appropriate Authority appointed under sub-section (1) of section 13; or

(ii) the State Government, where the appeal is against the order of the Authorisation Committee constituted under clause (b) of sub-section (4) of section 9 or against the order of the Appropriate Authority appointed under sub-section (2) of section 13.

CHAPTER VI

Offences and Penalties

18. (1) Any person who renders his services to or at any hospital and who, for purposes of transplantation, conducts, associates with, or helps in any manner in the removal of any human organ without authority, shall be punishable with imprisonment for a term which may extend to five years and with fine which may extend to ten thousand rupees.

(2) Where any person convicted under sub-section (1) is a registered medical practitioner, his name shall be reported by the Appropriate Authority to the respective State Medical Council for taking necessary action including the removal of his name from the register of the Council for a period of two years for the first offence and permanently for the subsequent offence.

19. Whoever—

(a) makes or receives any payment for the supply of, or for an offer to supply, any human organ;

(b) seeks to find a person willing to supply for payment any human organ;
(c) offers to supply any human organ for payment;
(d) initiates or negotiates any arrangement involving the making of any payment for the supply of, or for an offer to supply, any human organ;
(e) takes part in the management or control of a body of persons, whether a society, firm or company, whose activities consist of or include the initiation or negotiation of any arrangement referred to in clause (d); or
(f) publishes or distributes or causes to be published or distributed any advertisement, —

(a) inviting persons to supply for payment of any human organ;
(b) offering to supply any human organ for payment; or
(c) indicating that the advertiser is willing to initiate or negotiate any arrangement referred to in clause (d),

shall be punishable with imprisonment for a term which shall not be less than two years but which may extend to seven years and shall be liable to fine which shall not be less than ten thousand rupees but may extend to twenty thousand rupees:

Provided that the court may, for any adequate and special reason to be mentioned in the judgment, impose a sentence of imprisonment for a term of less than two years and a fine less than ten thousand rupees.

20. Whoever contravenes any provision of this Act or any rule made, or any condition of the registration granted, thereunder for which no punishment is separately provided in this Act, shall be punishable with imprisonment for a term which may extend to three years or with fine which may extend to five thousand rupees.

21. (1) Where any offence, punishable under this Act, has been committed by a company, every person who, at the time the offence was committed was in charge of, and was responsible to, the company for the conduct of the business of the company, as well as the company, shall be deemed to be guilty of the offence and shall be liable to be proceeded against and punished accordingly:

Provided that nothing contained in this sub-section shall render any such person liable to any punishment, if he proves that the offence was committed without his knowledge or that he had exercised all due diligence to prevent the commission of such offence.

(2) Notwithstanding anything contained in sub-section (1), where any offence punishable under this Act has been committed by a company and it is proved that the offence has been committed with the consent or connivance of, or is attributable to any neglect on the part of, any director, manager, secretary or other officer of the company, such director, manager, secretary or other officer shall also be deemed to be guilty of that offence and shall be liable to be proceeded against and punished accordingly.

Explanation.—For the purposes of this section,—

(a) "company" means any body corporate and includes a firm or other association of individuals; and
(b) "director", in relation to a firm, means a partner in the firm.

22. (1) No court shall take cognizance of an offence under this Act except on a complaint made by —
(a) the Appropriate Authority concerned, or any officer authorised in this behalf by the Central Government or the State Government or, as the case may be, the Appropriate Authority; or
(b) a person who has given notice of not less than sixty days, in such manner as may be prescribed, to the Appropriate Authority concerned, of the alleged offence and of his intention to make a complaint to the court.

(2) No court other than that of a Metropolitan Magistrate or a Judicial Magistrate of the first class shall try any offence punishable under this Act.

(3) Where a complaint has been made under clause (b) of sub-section (1), the court may, on demand by such person, direct the Appropriate Authority to make available copies of the relevant records in its possession to such person.

CHAPTER VII

Miscellaneous

23. (1) No suit, prosecution or other legal proceeding shall lie against any person for anything which is in good faith done or intended to be done in pursuance of the provisions of this Act.

(2) No suit or other legal proceeding shall lie against the Central Government or the State Government for any damage caused or likely to be caused for anything which is in good faith done or intended to be done in pursuance of the provisions of this Act.

24. (1) The Central Government may, by notification, make rules for carrying out the purposes of this Act.

(2) In particular, and without prejudice to the generality of the foregoing power, such rules may provide for all or any of the following matters, namely:

(a) the manner in which and the conditions subject to which any donor may authorise removal, before his death, of any human organ of his body under sub-section (1) of section 3;
(b) the form and the manner in which a brain-stem death is to be certified and the conditions and requirements which are to be satisfied for that purpose under sub-section (6) of section 3;
(c) the form and the manner in which any of the parents may give authority, in the case of brain-stem death of a minor, for the removal of any human organ under sub-section (7) of section 3;
(d) the form in which authority for the removal of any human organ from an unclaimed dead body may be given by the person incharge of the

management or control of the hospital or prison, under sub-section (1) of section 5;

(e) the steps to be taken for the preservation of the human organ removed from the body of any person, under section 7;

(f) the form and the manner in which an application may be jointly made by the donor and the recipient under sub-section (5) of section 9;

(g) the manner in which all possible effects, complications and hazards connected with the removal and transplantation is to be explained by the registered medical practitioner to the donor and the recipient under section 12;

(h) the standards as are to be enforced by the Appropriate Authority for hospitals engaged in the removal, storage or transplantation of any human organ under clause (iii) of sub-section (3) of section 13;

(i) the other measures as the Appropriate Authority shall undertake in performing its functions under clause (vi) of sub-section (3) of section 13;

(j) the form and the manner in which an application for registration shall be made and the fee which shall be accompanied, under sub-section (2) of section 14;

(k) the specialised services and the facilities to be provided, skilled manpower and the equipments to be possessed and the standards to be maintained by a hospital for registration, under sub-section (3) of section 14;

(l) the form in which, the period for which and the conditions subject to which certificate of registration is to be granted to a hospital, under sub-section (1) of section 15;

(m) the manner in which and the fee on payment of which certificate of registration is to be renewed under sub-section (3) of section 15;

(n) the manner in which an appeal may be preferred under section 17;

(o) the manner in which a person is required to give notice to the Appropriate Authority of the alleged offence and of his intention to make a complaint to the court, under clause (b) of sub-section (1) of section 22; and

(p) any other matter which is required to be, or may, be, prescribed.

(3) Every rule made under this Act shall be laid, as soon as may be after it is made, before each House of Parliament, while it is in session, for a total period of thirty days which may be comprised in one session or in two or more successive sessions, and if, before the expiry of the session immediately following the session or the successive sessions aforesaid, both Houses agree in making any modification in the rule or both Houses agree that the rule should not be made, the rule shall thereafter have effect only in such modified form or be of no effect, as the case may be; so, however, that any such modification or annulment shall be without prejudice to the validity of anything previously done under that rule.

25. (1) The Ear Drums and Ear Bones (Authority for Use for Therapeutic Purposes) Act, 1982 and the Eyes (Authority for Use for Therapeutic Purposes) Act, 1982 are hereby repealed.

(2) The repeal shall, however, not affect the previous operation of the Acts so repealed or anything duly done or suffered thereunder.

RULES

Ministry of Health and Family Welfare

G.S.R. 51 (E).—(New Delhi, the 4th February, 1995)—In exercise of the powers conferred by sub-section (1) of Section 24 of the Transplantation of Human Organs Act, 1994 (42 of 1994), the Central Government hereby makes the following rules, namely:

1. *Short title and commencement.*—(1) These rules may be called the Transplantation of Human Organs Rules, 1995.

(2) They shall come into force on the date of their publication in the Official Gazette.

2. *Definitions.*—(a) "Act" means the Transplantation of Human Organs Act, 1994 (42 of 1994);

(b) "Form" means a form annexed to these Rules;

(c) "Section" means a section of the Act;

(d) Words and expressions used and not defined in these Rules, but defined in the Act, shall have the same meanings respectively assigned to them in the Act.

3. *Authority for Removal of Human Organs.*—Any donor may authorise the removal, before his death, of any human organs of his body for therapeutic purpose in the manner and on such conditions as specified in Form 1.

4. *Duties of the Medical Practitioner.*—(1) A registered medical practitioner shall, before removing a human organ from the body of a donor before his death satisfy himself—

(a) that the donor has given his authorisation in Form 1;

(b) that the donor is in proper state of health and is fit to donate the organ, and shall sign a certificate as specified in Form 2;

(c) that the donor is a near relative of the recipient, and shall sign a certificate as specified in Form 3 after carrying out the following tests on the donor and the recipient, namely—

(i) tests for the antigenic products of the Human Major Histocompatibility system HLA-A, HLA-B and HLA-DR using conventional serological techniques;

(ii) tests to establish HLA-DR beta and HLA-DQ beta gene restriction fragment length polymorphism;

(iii) where the tests referred to in suo-clause (i) and sub-clause (ii) do not establish a genetic relationship between the donor and the recipient tests to establish DNA polymorphisms using at least two multi-locus gene probe;

(iv) where the tests referred to in sub-clause (iii) do not establish a genetic relationship between the donor and the recipient further tests to establish DNA polymorphisms using at least five single locus polymorphic probes.

(d) in case recipient is a spouse of the donor, record the statements of the recipient and the donor to the effect that, they are so related and shall sign a certificate in Form 4.

(2) A registered medical practitioner shall, before removing a human organ from the body of a person after his death satisfy himself—

(a) that the donor had, in the presence of two or more witnesses (at least one of whom is a near relative of such person), unequivocally authorised as specified in Form 5 before his death, the removal of the human organ of his body, after his death, for therapeutic purposes and there is no reason to believe that the donor had subsequently revoked the authority aforesaid;
(b) that the person lawfully in possession of the dead body has signed a certificate as specified in Form 6 or Form 7.

(3) A registered medical practitioner shall, before removing a human organ from the body of a person in the event of his brain-stem death, satisfy himself—

(a) that a certificate as specified in Form 8 has been signed by all the members of the Board of medical expert referred to in sub-section (6) of section 3 of the Act;
(b) that in the case of brain-stem death of a person of less than eighteen years of age, a certificate specified in Form 8 has been signed by all the members of the Board of medical experts referred to in sub-section (6) of section 3 of the Act and an authority as specified in Form 9 has been signed by either of the parents of such person.

5. *Preservation of organs.*—The organ removed shall be preserved according to current and accepted scientific method in order to ensure viability for the purpose of transplantation.

6. The donor and the recipient shall make jointly an application to grant approval for removal and transplantation of a human organ, to the Authorisation Committee as specified in Form 10.

7. *Registration of Hospital.*— (1) An application for registration shall be made to the Appropriate Authority as specified in Form 11. The application shall be accompanied by a fee of rupees one thousand payable to the Appropriate Authority by means of a bank draft or postal order.

(2) The Appropriate Authority shall, after holding an inquiry and after satisfying itself that the applicant has complied with all the requirements, grant a certificate of registration, as specified in Form 12 and shall be valid for a period of five years from the date of its issue and shall be renewable.

8. *Renewal of Registration.*—(1) An application for the renewal of a certificate of registration shall be made to the Appropriate Authority within a period of three months prior to the date of expiry of the original certificate of registration and shall be accompanied by a fee of rupees five hundred payable to the Appropriate Authority by means of a bank draft or a postal order.

(2) A renewal certificate of registration shall be as specified in Form 13 and shall be valid for a period of five years.

(3) If, after an inquiry including inspection of the hospital and scrutiny of its past performance and after giving an opportunity to the applicant, the Appropriate Authority is satisfied that the applicant, since grant of certificate of registration under sub-rule (2) of Rule 7 has not complied with the requirements of this Act and the Rules made thereunder and conditions subject to which the certificate of registration has been granted, shall, for reasons to be recorded in writing, refuse to grant renewal of the certificate of registration.

9. *Conditions for grant of Registration of Certificate.* — No hospital shall be granted a certificate of registration under this Act unless it fulfills the following requirements of manpower, equipment, specialised services and facilities.

[Omitted]

10. *Appeal.* — (1) Any person aggrieved by an order of the Authorisation Committee under subsection (6) of Section 9, or by an order of the Appropriate Authority under sub-section (2) of Section 15 and Section 16 of the Act, may, within thirty days from the date of receipt of the order prefer an appeal to the Central Government.

(2) Every appeal shall be in writing and shall be accompanied by copy of the order appealed against

[Forms Omitted]